a30118 021900497b

Yorkshire's Whaling Days

by

Jack Dykes

DALESMAN BOOKS
1980

THE DALESMAN PUBLISHING COMPANY LTD.,
CLAPHAM (via Lancaster), NORTH YORKSHIRE

First published 1980

ISBN: 0 85206 570 1

By the same author: SMUGGLING ON THE YORKSHIRE COAST

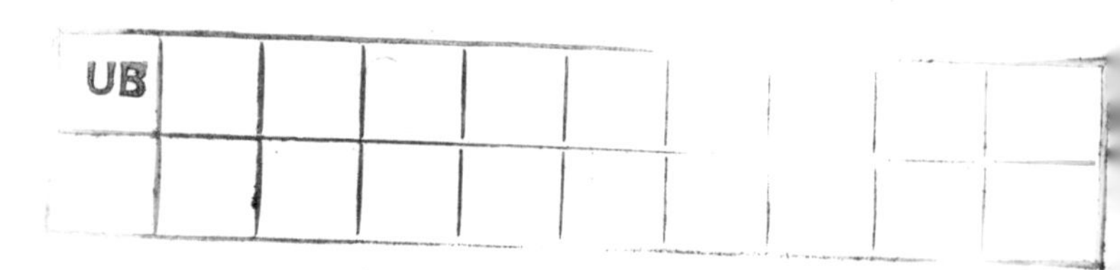

Printed in Great Britain by
GEORGE TODD & SON
Marlborough Street, Whitehaven

Contents

1	The Early Fishery	5
2	Return of the 'Esk'	10
3	Bear and Whale	14
4	Press-Gangs	27
5	The Greenland Men	30
6	The Wrecking of the 'Esk'	35
7	Ships and Ice	38
8	The 'Diana'	51
9	The End of the Trade	57
10	The End of the Whales	62

Illustrations in the text are on pages 17–24 and 41–48.

Chapter One

THE EARLY FISHERY

A WHALE is an intelligent, large-brained mammal, warm-blooded, and without gills, which needs to surface periodically to blow out stale air and mucus, and to recharge its vast lungs before diving.

Nevertheless, to the whalemen the creature was always a fish. When they harpooned one, and killed it with lances like long spears, which they stabbed through its skin and blubber, the captains sketched a whale's tail in the margin of the whaling log, and wrote alongside something like: 'At 10 a.m. this Forenoon a Fish caught in Calm Whether, rising near the edge of a Gigantic field of Ise. William Robson, Harpooner. Four boats Hemployed in towing the Fish back to the Ship.'

The particular species of 'fish' hunted in the Arctic was the 60′–0″ giant known as the Greenland Right Whale, a plankton-eating, mild, and inoffensive animal, with but two enemies: the pack-hunting killer-whales, and, with a less dramatic name, but nevertheless far more deadly, man, who has succeeded in decimating the creature to the point of extinction. Other sea-dwellers that the whalemen hunted, generally as alternatives if the whales were scarce, were the grey-white beluga, a mere 18′–0″ long, which they called the 'sea-canary'; the narwhal, 16′–0″ long, called the 'unicorn' because of its long, spiral, tusk; the walrus; and especially the seals of Greenland and Newfoundland, best killed when babies, still unable to swim and wearing the fine white coats, with inch thick fur, so prized by women of fashion.

The voyage of a whaler was, or was always intended to be at the outset, that of a ruthless, implacable, and invincible, hunter. Nothing of flesh and blood that the whale-ship might obtain was safe. Not the whales, the seals, the walrus, the polar bears or the arctic foxes. Not even guillemots, auks, burgomaster gulls, nor snowy owls. What was not caught for profit was slaughtered either for the pot or for 'sport.' In the early 19th century heyday of whaling, the ships left Humber and Esk in February or March, made up their complement of seamen at Orkney or Shetland, and then proceeded to stain the Arctic ice and sea with the blood and bones of whatever they could find.

It was a pitiless trade, but the slaughter was localised, and diminished in the vastness of the northern wilderness, and, if thought about at all, was justified by a recounting of the risks taken, for the whaling was an annual dicing with sub-zero temperatures, shifting ice, hunger, and disease.

Almost every year in the busy decades, ships failed to return, and were either never seen or heard of again, or else limped back across the Atlantic with a dead or dying crew. Horrific stories abound, documented by eye-witness accounts. Perhaps only a single ship would be concerned, trapped by the ice and isolated in the high latitudes, as removed from all human help as effectively as if on another planet; or perhaps there would be many, as in 1820, when, with 11 ships from Whitby, and 62 from Hull, so many vessels in the combined English-Scottish fleet were beset and threatened by shifting ice fields in Baffin Bay, that hundreds of men took to the ice, lit fires, and got at the rum, to create an event known ever after as 'Baffin Fair.'

Some owners provisioned their ships better than others, but even at the very end of the trade some Hull owners were stocking their whale-ships for a bare eight months' voyage, when time after time ships trapped in the ice had arrived back months late—perhaps not until the ice had released them in early spring, so that at the very time when men needed extra nutrition to withstand extreme cold, they were in fact reduced to unsuitable rations scarcely sufficient to ward off starvation.

Indeed, it is arguable as to where the whalemen's greatest dangers came from, the foul Arctic weather, or the owners, for not only was the food—basically salt-meat and ship's biscuit—often limited in quantity, but there was a callous disregard concerning scurvy. The navy had realised the effectiveness of lime and lemon juice since the mid-18th century, yet the mercantile marine had to be compelled, by the Merchant Shipping Act of 1854, to carry anti-scorbutics on all voyages of more than 10 days. As late as 1866 a Hull whaler, ignoring these Board of Trade regulations, put to sea with scarcely a token quantity; the result being, when she was trapped by ice, that men died slowly and painfully of the disease, which, caused by a deficiency of ascorbic acid, is easily reversible if vitamin C can be obtained. If not, death is certain.

There was never a trade so colourful, so crude, so fulfilling, and so destroying. It was a calling of contrasts; a ship returning full of whale-bone and blubber meant wealth for all, even the crew, whose low wages would be supplemented by a bonus; whilst a ship returning empty, or, in the whalers' jargon, 'clean,' meant absolute loss for the owners and often debt for the crew, who would have drawn advances on payments which had not materialised. Again, a ship would leave Hull or Whitby with its seams newly caulked, its rigging tarred, its masts and spars oiled, and perhaps with unblemished new canvas. Crowds of relatives and friends and well-wishers waved and cheered to men and boys hearty and full of hope, and it was customary for a garland made from the ribbons of wives and sweethearts to be lashed a hundreed feet above deck, near the top of the mainmast; yet six months later, there might still be no news of that ship, which might eventually return defaced and battered, with some of the crew dead or missing, and most of the remainder sick and emaciated.

Whaling from Whitby began in 1753, when the *Henry and Mary* and the *Sea Nymph* sailed for Greenland, and lasted until 1837.

In the peak years the town had proportionately more people involved in the trade than any other place in Britain, including Hull. At the first national census of 1801, the population of Whitby was 7,500, that of Hull, 29,500; the greatest number of whaleships to leave Whitby in one season was 20, in each of the three years 1786, 1787, and 1788; in Hull, 63. Each ship would have at least 25 local men and boys, and up to twice that number if no Orkney or Shetland men were to be signed on. A few of the ships were Whitby built. And all of them had to be equipped and provisioned, with, for example: salted pork and beef, potatoes and flour, ship's biscuit, tobacco, tea and sugar, timber, canvas, rope, tools, casks, and candles. The blubber that they brought back had to be rendered into oil, the baleen, or whalebone, had to be cleaned, cut and shaped, and finally all the produce had to be taken to the 'manufactories,' either by coaster, or by horse and cart.

Numerous, unrecorded, captains brought their ships back from Greenland or Davis Straits in late summer or in autumn, sailing over the bar at flood tide, and between the leaves of the open drawbridge into the inner harbour, and a noisy welcome. Their ships were still greasy, despite the cleaning, and most of the men were burned brown and looked as if they had spent months shearing sheep up on the moors. Crowds would gather, men and boys would be anxious to get off to wives and sweethearts, homes and pubs, and the longshoremen would descend upon the ships to remove the casks of oozing blubber, the boiling of which, in four oil yards, two at each side of the river, would soon have the town stinking with a peculiar, oppressive, and lingering, smell.

Sixty miles further south, Hull had always been one of the major ports in the country, and at the end of the 16th century shared with Lynn and with London the distinction of being concerned in the very first English whaling, around the coast of Iceland and north Norway, and especially at Spitsbergen.

The islands of Spitsbergen lie well to the north of Norway, between 76 degrees and 80 degrees of latitude, in a region which has total daylight for 122 days each year, and no sun at all for 115. The islands form a land of contrasts, in summer with remarkable visibility in the crystal clear air, and with snow and ice sparkling under hazy blue skies; and in winter with dense fogs, hurricanes, appalling cold, perpetual twilight and darkness. For long periods of time, whales or different species had inhabited Spitsbergen's fjords and shallow bays within the natural order of life, until man arrived with the virulence of bubonic plague.

For the whales, it was the beginning of the end when those first English ships nosed around the coast, noting with relish the size, the numbers, and the tameness, of the creatures. Here was an El Dorado of animal flesh, easily slaughtered, and with high profits. Pure greed, however, caused the English to squabble, firstly with each other, and then, when other nations had become interested—the energetic Dutch, the French, the Spanish, with their years of experience of whaling in the Bay of Biscay, the Germans, and the Norwegians—with the people they regarded as interlopers. Soon, the English and the Dutch were threatening each other with gun boats. In 1613, the London Russia Company, which had been granted monopoly rights by the king, sent armed ships and drove away the Dutch fleet; and in the following year the Dutch were accompanied by four 30-gun ships, which drove away the English fleet.

Even with the Dutch as gunboat competitors, the English merchants were unable to agree amongst themselves, for the London company continually harassed the Hull men. The captain of one London whaler complained that when he met up with a Hull ship hunting whale and walrus, the Yorkshireman refused to go, 'saying that there was enough for all, and that we durst do nothing, for he had a number of stout North Countrymen aboard who were the equal and more to any of ours.'

In 1622 the mayor and aldermen of Hull made a lengthy submission about the falling trade of the town to the Privy Council in London, part of which dealt with whaling: '. . . we did seek to revive again, by searching and finding out the land called Greenland, where we were the first that found that country, and gave the first hazard of any *Englishmen,* to kill the whale, which voyage we hoped would be the setting to work a great quantity of poor people and artificers of this Town . . . but the Company of *Russia* merchants in *London,* do so exceedingly disturb us therein, that we humbly desire your Honours favour and redress.' The following year a fleet of nine whalers, funded from Hull and York, first visited the Spitsbergen headquarters of the London fleet, demolished all the buildings, broke the iron coolers, and set fire to what was left.

Eventually, as the English harassed each other, and the Dutch exploited the abundance of whales and seals and bears with great energy, national areas of fishing rights were established, and ships from Hull continued to search for the already diminishing numbers of whales, which, in desperation, were moving away from open water and into the ice.

George Hadley, who published in 1788 a history of Hull 'compiled from official records,' detailed a dramatic incident in 1631. He referred to the land as Greenland, but it seems likely to have been Spitsbergen.

In this year an English whaleship, responding to unexpected and frantic signals, came upon eight thin, dirty, dishevelled men, scarcely recognisable as civilised human beings, and was told that the previous year they had landed, by boat, from the *Salutation,* out of Hull, to hunt for deer, when fog and high winds parted them from their ship. They had with them, 'a brace of dogs, a firelock, two lances, and a tinder box.' Delayed by shifting ice, and by having to drag their boat over land, they were late at a pre-determined rendezvous, and found themselves stranded, with winter, which begins in September, a few short weeks away. They killed 19 deer and four bears, a remarkable achievement if in fact they had but the one flintlock gun, and rowed along the coast to a place where visiting ships had boiled out blubber, and here they erected the heavy canvas tent they had in the boat, gathered bits of wood as fuel, killed two walrus, ceased to wash, and slept wrapped up in the deer skins.

Food was soon short—even bears grow lean and emaciated in an Arctic winter—and for days on end the weather was so awful that they could scarcely go out of the tent, now held down by stones and rope. Clothes became greasy and filthy, chaffing the skin around wrists and neck. Soon, they had no food at all for two days each week, and on the others helped out their diminishing stock of venison and bear with mouldy scraps of whale, left by the whale ships as of no use. The cold was almost unbearable; there was so little fuel that they could afford a fire only to take the iron hardness from the frozen meat; and they had nothing to drink but tepid water melted from snow.

By February it was obvious that they would all die if they could not find more food, and, almost at the end, they were fortunate enough to kill a she-bear, approaching with her cub. By March the weather was improving, and they trapped the returning birds, and several foxes. And in May, more welcome even than the sun, the first of the English ships appeared.

For a hundred years and more, English whaling languished, whilst that of the Dutch became a national enterprise, and Hadley calculated that in the 46 years ending in 1720 the Dutch had sent 150 ships yearly, each killing on average five whales. These figures are suspect, but there is no doubt that so many whales had been slaughtered by the time the English again became interested, that the Spitsbergen grounds had become uneconomic, and the whalers were moving towards the east coast of Greenland, and eventually, to Davis Straits and Baffin Bay.

With increasing oil demand for industry, whalebone for corsets and umbrellas, and seal skins for clothing and fashion, Hull re-commenced whaling a few years after Whitby began. The number of vessels involved multiplied rapidly, and by the late 18th century, Hull had over 30 vessels and was second only to London, and Whitby had between 10 and 20, and was running either third in the country, or fourth, behind booming Liverpool. In the second last decade the town of Stockton, at this time the most important port on the Tees, had a couple of vessels, and Scarborough completed the Yorkshire fleet with a single ship.

Chapter Two

RETURN OF THE 'ESK'

THERE was rather more to Arctic whaling than an expedition being decided upon, a ship fitted, a crew engaged, and then the whales being caught or, at least, chased. Much depended upon luck, and upon the unveiling of factors unknown when the ship sailed, such as the quantity of ice and the direction and force of the winds. Much more depended upon the captain of the vessel.

When it came to the welfare of their crews, there were good captains and bad ones, but there were very few inefficient or frightened ones, for if a man did not show a reasonable return within one or two seasons, then he was out of a job. Whaling could make a captain wealthy, but it forced upon him a tremendous responsibility, not only to the owners, but to every man and boy of the crew, and sometimes even to the crews of other ships.

When a whaleship left Britain for the Greenland Sea, or for Baffin Bay, the captain virtually had absolute power on his floating island of wood and sail, and, in a crisis, his decisions could mean the difference between life and death, and between capitulation and a refusal to accept what seemed inevitable. The voyage of the *Esk,* of Whitby, in 1816, illustrates this neatly.

The captain was William Scoresby, junior, aged 26, a local man, son of a famous whaling father, and himself an unusual combination of whaling captain and scientist, already collecting material for his superlative two-volume work on whaling and the Arctic, which was published in 1820.

The *Esk* had caught and flensed several good-sized whales, although there was much ice about, and, by the end of June, Scoresby was in the company of several other ships, English and Scottish, roaming the Greenland Sea as high as 78 degrees of latitude, seven or eight hundred miles north of Iceland, and amongst great floes and fields of ice.

On June 28th, with the rest of the fleet in open water, Scoresby took his ship actually into the pack ice, a risky thing to do but a manoeuvre which had paid off in the past, for if whales were found in the lanes and large pools of open water, then there was no one else to claim them. This time, however, there were no whales and the ice began to move, the lanes closing inexorably. Scoresby, experienced in such a situation, was not unduly worried, and ordered the whale boats to fix a rope tow to the *Esk* and row her back to open water. Yet the closing ice seemed to increase its pace, and before long Scoresby had to make a hasty turn into a harbour within the field.

The *Esk* was now actually beset within the diminishing area of water into which the ice was moving, with colossal pressures building up, to a degree which could crush the whaler as if she was an egg-shell; but by gauging the drift, and the strength, of the encroaching ice, a captain could ease his ship into comparative safety, allowing the main pressure forces to flow by and around him, as if the ship was a fish sheltering in the eddy of a rock.

Large iron hooks, called 'ice anchors,' were driven into the ice to act as warping posts, and were used to manoeuvre the ship into a small bay surrounded by thick ice. Although the water disappeared, and the pressure was enough to lift the ship's heavy wood and iron rudder, weighing several tons, and although the *Esk* shivered and groaned, there seemed to be nothing more than that. Outside the field of ice the other whalers watched with interest, for it was not unusual for a nipped ship, with her bottom torn off, to appear quite undamaged, until the ice parted and she sank like a stone.

After some hours, with the weather now misty, damp, and cold, the pressure eased, the ice drew back, and, as a normal precaution, the carpenter sounded the pumps. To his horror, they drew deep water, and a quick visual inspection showed that there was over eight feet in the hold, a quantity which Scoresby later calculated to have weighed almost 200 tons, about half the weight of the ship. What had happened was that a submerged tongue of ice had pushed against the keel and torn loose over 20 feet of the after part of it, together with a plank. Water was pouring in, and Scoresby hoisted the distress signal. There was a quick response from the nearest ships, which were some hundreds of yards away.

The *Esk* had two pumps of 9″ bore, and one of 6″, cumbersome wood and leather contraptions, operated by hand, and Scoresby soon had men from the crews of half a dozen ships, 150 in all, pumping and baling. When the water level in the hold was reduced to 4′–0″, bundles of oakum and sail were introduced into the long, horizontal, hole, with the idea of letting the flow of incoming sea wedge them in, but they had little effect. Pumping and baling continued for 40 solid hours, until the *Esk's* own crew was completely exhausted, and the seamen from the other ships, who figured they were working too hard for nothing, had had enough. At no time had they been able to suck the hold dry, and immediately the pumps stopped the water gained rapidly.

Scoresby now decided that they must expose the bottom of the ship, in order to assess the full extent of the damage, and to get at it properly. However, though it was essential that no time be wasted, he had now gone 50 hours without rest, and his body was giving way. Two tents were erected on the ice, and captain and crew slept—for four hours. Rope hawsers were then fixed from the masts, under the ship, and to the ice, and the anchors were hung from fore and main mast heads, to aid, by their weight, the canting over of the ship. Yet even with 150 men heaving on the tackles, the ship could not be heeled far enough to expose the damage, which was at the very base of the hull. The seamen from the other ships now left, tired out, hungry, and fatalistic about the *Esk's* chances. By now, the Whitby men were literally reeling with exhaustion.

An attempt was made to rope and rip loose completely the damaged keel, and this was successful. A sail, covered with oakum and old canvas, was passed over the hole and lashed into place, the concavity of it padded and filled with more oakum, and another sail fitted over that. The crew of the *John,* a Greenock ship, whose captain was Scoresby's brother-in-law, returned, and the extremely arduous and tedious task of pumping began again. With the water diminished, the carpenter of the *Esk* and two others attempted to repair the damage from inside, with oakum, wood, pieces of fat pork, and tar. They had only limited success, for conditions were almost intolerable, dark and cramped, and with the men's arms and legs, and sometimes most of their bodies, being continually submerged in near-freezing water. Despondency was now rife, and noticeable, with the men beginning to soldier, and to skylark in a desultory manner. Scoresby learned that the crew of the *John* were on the verge of quitting, and that most of his own crew intended to go with them.

A hasty discussion took place between Captain Scoresby and Captain Jackson of the *John,* and then with the respective crews, who stood to lose or gain a bonus. It was agreed that the Greenock ship should stay with the *Esk* and give her every assistance, towing her to within reach of a Shetland port, in return for 48 tons of whale oil and 1½ tons of whalebone, which was almost half the *Esk's* catch, plus also £100 for the seamen. A formal agreement was drawn up, and when the ice opened sufficiently, the *John* came alongside, 130 butts of blubber were transferred—and both captains were compelled to rest, Scoresby having had only 12 hours sleep in the last five days, and his legs now so swollen that he could scarcely move.

When the *John* began eventually to tow the *Esk,* it was discovered that the damage and the repairs to her hull had made her unmanageable, her rudder having no effect at all on her steering. This was almost the final blow, damaging to the spirit of even the few optimists left in the Whitby ship. The weather was bad, the wind rising, and the ice moving. They were thoroughly exhausted, and were several hundred miles away from even the desolate north coast of Iceland. All ships other than the *John* had left them, and the *Esk* was almost a wreck. Nevertheless, with blocks and tackle they managed to wrestle the rudder onto the ice and add another 20 square feet to its area. Replaced, it made the ship just about navigable, rather better than an immense log. It took two days of hard work for the Greenock ship to tow her clear of the ice.

With a heavy rope spring between the two whalers, the *John* now began the thousand mile, open-water, tow, sailing due south, in seas which, even in the height of summer, can provide some of the world's foulest weather. Fortunately the wind was mainly from the north, and crashed the square-riggers along for mile after mile at nine knots. On board the *Esk* there was little rest, for water was flowing into the hold even when the packing braced over it was secure, and twice during the hair-raising trip the heavy seas tore the outer canvas off the plunging ship. When this happened, there were frantic signals to the *John* to ease up, before the Whitby ship filled and foundered.

The last 300 miles of the tow took seven days and nights, with the *Esk,* in bad weather, scarcely more manageable than a piece of the ice which had holed her. Nevertheless, on the 27th of July the northernmost part of Shetland, the island of Unst, was seen as a low, solid, cloud on the southern horizon, and later that day the two whalers parted company, and the *Esk* lumbered safely into Lerwick harbour.

At Lerwick the ship was repaired, and Scoresby took her back into Whitby with at least a moderate catch, and with the crew alive and well.

It could so easily have been different; no ship and consequently no catch, and the crew dispersed amongst other vessels; or the towed ship might have foundered in heavy seas, and so no ship and probably considerably less crew.

Whatever money the whalemen made, they earned.

Chapter Three

BEAR AND WHALE

THE Arctic had a strange and terrible beauty about it, and perhaps still has, although nowadays more difficult to appreciate from the towering steel deck of a 140,000 ton super tanker, battering through the ice of Baffin Bay to Alaska, than from the wooden planking of a 350 ton, three masted sailing ship, with a limited supply of food and no chance of any help if she became fast. Several captains, notably William Scoresby, junior, from Whitby, in the early part of the 19th century, and William Barron, from Hull, in the middle part, described in print, and with fluency, this fantastic panorama of sea and ice and hazy sky, the sculptored icebergs, the colossal cliffs of the islands and of Greenland, the shifting ice fields, the Northern Lights, and the midnight sun.

At that time, most people in England thought that the Arctic was a bleak, black and white, wilderness, displaying all the worst attributes of a raw and miserable English winter, yet of course this was not so. Away from land the icebergs often had in them translucent shades of blue and green and yellow, with miniature waterfalls glistening in the sunlight, and with dark caves and grottoes formed in the melting ice. On land, in the summer, there were patches of low-lying flowers: saxifrage, stitchwort, dandelions, cinquefoil—often unexpected, and therefore doubly delightful—occasional crops of crowberries, bilberries, and cranberries, and varying shades of moss and lichen. In winter, if a ship was unfortunate enough to become fast, there were incredible icescapes; night skies of a clarity impossible in Britain; and shimmering displays of the Aurora Borealis, known by the Shetlanders as the 'Merry Dancers.'

Ships' surgeons, who were often the only formally educated men on the ships, liked to keep journals of their voyages. Dr. Charles Edward Smith, surgeon of the ill-fated *Diana,* out of Hull in 1866, kept a comprehensive record, and wrote of this savage loveliness on January 25th, 1867. During the summer, on June the 12th, he had noted 'hot, calm, weather continues, men burned brown, dazzling light.' Now, beset in Baffin Bay, he wrote: 'Extremely cold, clear, weather, with glorious sunrises and sunsets lightening the whole sky magnificently, the colours creeping in the morning from the glowing sun, and bathing ice and snow in sparkling suffusions of orange, red, and yellow shadings. Much refraction, land, bergs, and islands, elevated and distorted, constantly assuming different shapes . . . scurvy cases increasing . . . yesterday afternoon was extremely cold, and the sea was giving out dense columns of thick, dark, vapour.'

Probably the two most impressive creatures of these regions—the whalemen referred to the whole area of land and sea and ice as 'the country,' to differentiate between it and the Atlantic—were the whale and the polar bear. The latter was called 'the Farmer' by the Yorkshire men, a nickname given because of the animal's powerful, lumbering, gait. The largest of all carnivores, with a diet composed almost solely of seals, it would often approach whalemen on shore, and they would generally attempt to kill the creature, mainly for its fur, for although the meat was eaten by hungry men, the flesh often contained the trichinosis parasite, which could prove fatal, and the liver had such a high concentration of vitamin A—which of course was not known to the 18th and 19th century whalers—that it was poisonous.

In the early part of the season she-bears would often be accompanied by a cub, and here the action would be to kill the mother and capture the young animal. The cub was invariably more puzzled and frightened than vicious, prodding its dead mother, sniffing at her, and making pathetic little mewing sounds, and it would be quickly trussed up and carried back to the ship, sometimes to be kept on deck in a barrel with bars fixed over the end, occasionally being dowsed with salt water, and fed with whale-meat. Back in England the animals would be sold to fairs and zoos, and Hull seems to have been well supplied with bears and seals at the town's Botanic Gardens. Life, hard and pitiless towards people, was even more so to the animals. Nevertheless, William Scoresby, senior, the most successful of the late 18th century whaling captains, proved that the animals responded to decent treatment.

He captured a cub off Greenland and fastened it to a point in the deck, training it by striking it firmly and painfully on its black nose if it attempted to bite his hand. By the end of the season he could lead it on a length of rope almost like a dog, but, back in Whitby, the animal, confused by the noises and movement and smells, broke free from the ship and ran into the town. Eventually, the bear was cornered in Cockmill Wood, by a great mass of men and women and children, with guns and pitchforks and dogs.

Fortunately, Captain Scoresby arrived before anyone was hurt or the animal killed, and added to his reputation by pushing through the crowd, walking up to the bear—which licked his hand with its long, rough, tongue—tying a length of rope around its neck, and leading it away. Soon afterwards the animal was taken to London's Tower Zoo.

Polar bears were killed by rifles; whales were hooked by the barbs of a harpoon, and then stabbed to death by the long, razor-sharp, lances of the harpooner. Throughout the period of Yorkshire whaling, experiments were made with harpoon guns, and whaleboats sometimes had them fitted on a swivel in the bows, but they tended to misfire, and were inaccurate, and were not perfected until after the Hull and Whitby trades had finished.

Hand harpoons were heavy weapons, as they had to be to penetrate the flesh and blubber of the whale, and then hold and take the strain as the barbs lodged in the fibrous muscle tissue and the harpoon line paid out. The harpooners had to be strong and lithe, knowledgeable in where to strike and then skilled enough to get in close and do it.

The first indication of a whale came from the look-out, perched in a barrel at the top of the mainmast. The best whaling weather was for the sea to be calm and the atmosphere clear, so that the black bulk of the whale could be easily seen when the animal sounded, that is, blew out the foul air from the double blowhole on the top of its head, and took in more air before diving again. It would generally be on the surface for not more than two minutes, and in that time, if no boats were already at sea, the crew had to leave whatever they were doing—grab clothes if they had been asleep—scramble into the boats, launch them, and row to the whale, or to where the harpooner calculated the creature would rise again.

There were usually six men to a double-ended whaleboat; four oarsmen, a boat-steerer aft, and a harpooner for'ard. Instead of a rudder, the boat-steerer had a long oar, which gave less water resistance, and the boats skimmed over the sea. The Orkney and Shetland men, and especially the latter, were reckoned to be the finest boatmen in the country, and the oars dipped in unison, with a long, powerful stroke, as the steerer urged his 'lads' to give another pound, 'Come *on,* my bonny lads, pull! Pull!'

In the bows the harpooner, invariably a Yorkshireman in the Hull and Whitby vessels, would be taking the balance of his heavy metal dart, with its murderous arrow-head barbs. When they reached the whale, the steerer's ploy was to approach from behind, so that the harpooner might strike in a certain vulnerable part slightly behind what approximated to the animal's neck. The instant that the harpoon, effective at about eight yards, was thrown, so the steerer had to have the boat out of the way, for the whale's immediate reaction was to pound the sea with its 20 foot wide tail flukes, which could smash any boat like so much matchwood.

When struck, the whale, terrified and in pain, would arch its back and dive with such rapidity that the harpoon line, running in a single loop around a bollard in the boat, had to be continually doused with buckets of water to avoid catching fire. If the sea was comparatively shallow, then the whale sometimes struck the bed with such force that it smashed the bones of jaw and head. It might roll, in blind panic, and become entangled in the rope, drowning itself; or, more likely, it would take out line after line. Each boat carried six lines of 2¼" circumference, each 120 fathoms long, and as one was paid out so another would be rapidly fastened to it. The strength of some whales was phenomenal; whilst in the *Baffin,* Captain Scoresby, junior, caught a whale which dragged two boats on the surface for some distance, to find that the creature was also dragging underwater six full-length lines, and also a submerged whaleboat belonging to a Hull ship.

Captain William Gravill, of Hull, stated that he had known whales stay underwater for two hours, but this was exceptional, and the creatures normally had to surface for air long before that. It was then, if the harpoon had held, that the whalemen strove to get near enough for the harpooners to plunge in their long-handled lances, fatal for the whale if the lances could be churned into the creature's vital

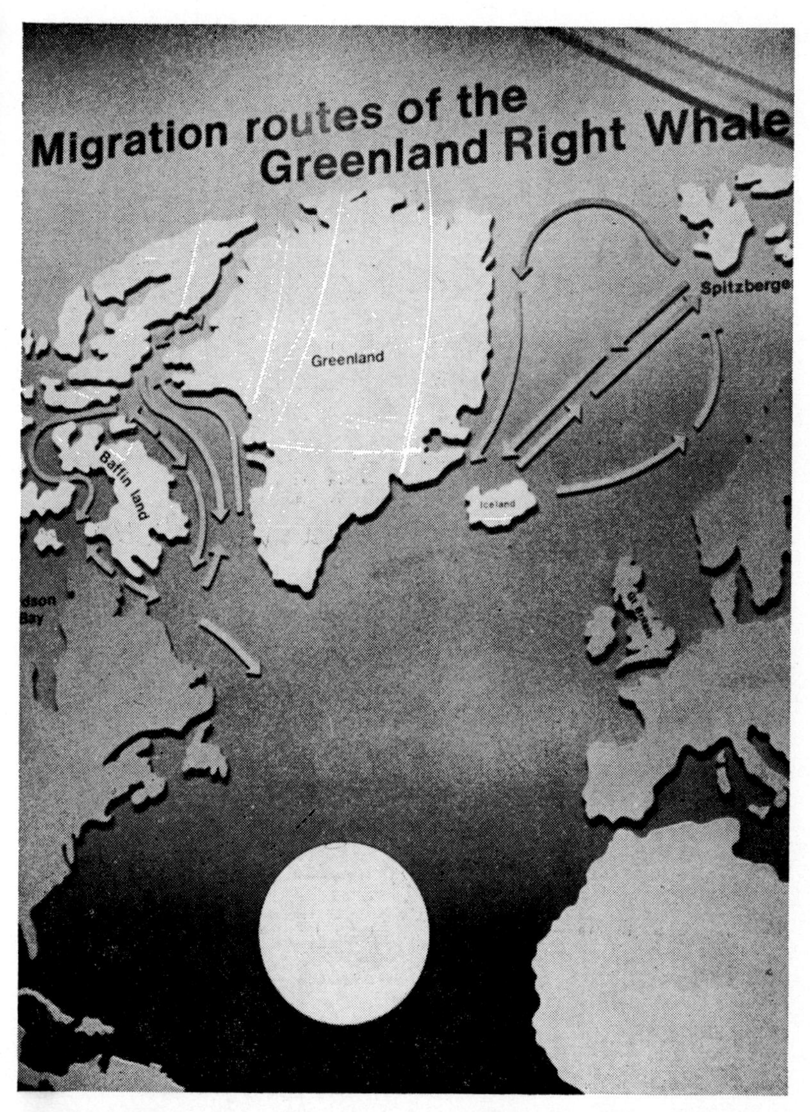

Arctic fishing grounds of the Yorkshire whalers. The distance between England's East Coast and the south tip of Greenland was at one time rumoured to be two thousand miles.

A CORRECT STATEMENT

OF THE

Success of the Hull Ships at the Greenland and Davis' Straits Fisheries,

In the Year 1821.

GREENLAND.

Ships' Names.	Register Tonnage. tons	pts.	No. of Men.	Captains' Names.	Date of arrival at Hull.	No. of Fish.	Actual weight of Fins or Whale Bone. tons.	cwt.	qr.	lb.	Actual quantity of Oil boiled. tons.	qr.	gall.
North Briton......	262	,,	41	John Allen......	Aug. 2	10	6	16	0	7	167	3	29
Perseverance......	251	,,	47	Matthew Wilburne..	— 10	10	8	6	3	13	151	1	9
Everthorpe	349	,,	47	Robert Ash	— 16	11	9	19	1	12	182	1	30
Cicero	325	5	41	William Leaf......	— 21	7	4	13	2	1	101	1	47
Mercury	346	25	48	William Jackson ..	— 21	13	7	14	3	20	161	3	44
Elizabeth	321	,,	45	Thomas Rhoades ...	— 22	10	5	16	2	20	131	0	36
Truelove	293	70	43	Thomas Todd	— 23	3	2	17	1	7	54	2	6
Walker...........	335	,,	49	Richard Harrison ..	— 25	7	5	14	0	26	105	0	24
Laurel	321	,,	41	Edward Dannatt ...	— 25	7	4	5	1	7	95	0	34
Manchester	285	,,	46	John Mitchinson ...	— 25	5	5	16	1	0	34	2	59
Shannon	348	,,	41	Robert Kelah	— 25	4	2	11	3	26	5	3	33
Ebor.............	283	23	49	Thomas Lee	— 26	5	4	8	0	11	8	3	35
Venerable	328	,,	41	John Bennet	— 27	10	4	17	0	17	[illegible]	0	4
William Torr......	280	80	46	Phillip Dannat	— 27	5	3	2	0	16	64	2	57
Dordon	285	90	49	William Gilyott	— 27	4	2	14	1	5	51	0	22
Neptune	336	17	49	Martin Munroe	— 30	5	4	2	1	3	[illegible]	3	60
Alfred	303	,,	49	William Clark	— 30		*Clean.*						
Gardiner and Joseph	360	,,	41	James Angus	— 31	7	2	19	3	19	84	1	38
Exmouth	321	21	41	Edward Thompson ..	— 31	1		2	2	18	3	0	24
Cyrus...........	346	48	42	William Beadling ..	— 31	7	4	9	1	22	9	0	11
Mary and Elizabeth	347	27	34	Robert Williams....	Sept. 1	1		6	2	11	1	2	51
Abram	319	14	48	William Harrison ..	— 1	3	2	13	0	8	54	0	55
Trafalgar	330	,,	46	William Lloyd......	— 1	7	4	3	1	3	98	3	41
Duncombe........	270	,,	45	John Corbett	— 13	9	6	3	3	2	136	3	4
Eagle............	289	,,	49	William Brewis....	— 13	14	7	13	1	3	154	0	58
Harmony	300	,,	45	Charles Sawyer....	— 15	3	2	10	2	27	52	2	24
Fame.............	377	12	56	William Scoresby ..	— 19	9	6	12	3	22	142	3	15
Jane	359	,,	41	Stephen Gamblin ...	— 20	1		18	3	9	17	2	37
Cato.............	305	,,	41	Andrew Turnbull...	— 20	1	*No Bone.*				5	1	46
Unity	272	81	35	William Short	— 22	11	5	18	3	21	117	2	54
Rachel and Ann ...	223	84	40	Richard Marshall ..	— 25	14	7	0	3	23	144	3	28
Total. 31 Ships	9565	33	1376			204	135	10	3	15	2745	2	7
Average each Ship	311	73 26/31	44 12/31			6 18/31	4	7	1	22 5/31	88	2	16 11/31

DAVIS' STRAITS.

Ships' Names.	Register Tonnage. tons	pts.	No. of Men.	Captains' Names.	Date of arrival at Hull.	No. of Fish.	Actual weight of Fins or Whale Bone. tons.	cwt.	qr.	lb.	Actual quantity of Oil boiled. tons.	qr.	gall.
Ellison...........	357	61	47	John Johnson	Sept. 19	13	10	15	0	15	170	3	56
Cumbrian	373	34	55	John Johnson	Oct. 4	28	12	14	1	25	213	3	34
Zephyr	342	,,	41	John Unthank	— 6	19	14	0	0	0	213	2	62
Gilder	360	7	41	George Bruce	— 6	19	12	19	0	23	219	0	33
Brunswick	357	,,	62	William Blyth	— 7	24	14	15	3	12	269	0	30
Andrew Marvel....	377	,,	41	Thomas Orton.	— 13	18	13	1	1	15	215	2	61
Albion	321	61	48	Richard Humphreys	— 17	21	12	15	1	7	214	1	39
Kirk Ella.........	410	72	49	Henry Watson	— 17	6	5	1	3	13	80	1	57
William	350	,,	46	Thomas Hawkins...	— 17	12	7	15	3	10	129	1	47
Lee	363	,,	41	Thomas Forster	— 17	11	8	14	2	20	123	2	4
Kiero.............	358	,,	46	James Colquhoun...	— 17	4	3	10	3	20	53	1	24
Eggiston	336	27	47	John Wilson	— 17	8	6	17	2	4	114	2	17
Lord Wellington ...	354	,,	41	John Boydon	— 18	15	10	4	3	26	144	1	52
Progress	307	56	49	Matthew Mercer....	— 18	8	5	5	3	16	97	2	57
Mary Frances	385	73	41	Thomas Wilkinson..	— 18	14	8	4	2	6	161	0	21
Ingria	318	,,	49	James Mackintosh..	— 19	20	[illegible]	8	1	14	[illegible]	[illegible]	[illegible]
Royal George	366	28	49	Joseph Peckit	— 20	9	5	9	1	19	108	0	0
Friendship........	410	,,	49	George Green......	— 21	6	4	1	2	21	80	0	15
Ariel.............	340	,,	49	William Hurst.....	— 21	12	6	7	0	15	118	0	23
Thomas...........	355	,,	41	William Brass	Nov. 6	10	7	10	1	7	128	3	1
Margaret	339	12	41	James Creighton ...	— 6	8	4	4	3	4	76	2	46
Total. 21 Ships	7482	55	963			294	186	19	1	12	3142	3	6
Average each Ship	356	29 19/21	45 18/21			14	8	18	0	7 5/21	149	2	39 6/21

GRAND TOTAL AND AVERAGE FOR GREENLAND AND DAVIS' STRAITS.

	Register Tonnage. tons	pts.	No. of Men.			No. of Fish.	tons.	cwt.	qr.	lb.	tons.	qr.	gall.
Total. 52 Ships	17147	88	2339			498	322	10	0	27	5888	1	13
Average each Ship	329	72 [illegible]	44 [illegible]			9 [illegible]	6	4	0	4 [illegible]	113	0	59 [illegible]

☞ The Refuse, or Black Oil, is not included in this Account, nor the number of Men taken from Shetland and the Orkneys.—The John, 343 Tons; Symmetry, 342 Tons; Harmony, 378 Tons; Leviathan, 409 Tons; Henry, 314 Tons; Cervantes, 355 Tons; Aurora, 368 Tons, Lost at Davis' Straits, Crews saved.—Thornton, 262 Tons, Lost at Greenland.—Hebe, 364 Tons, Lost on her passage to Davis' Straits, Crews saved.

AGGREGATE STATEMENT OF THE NUMBER OF THE HULL SHIPS FROM THE GREENLAND AND DAVIS' STRAITS FISHERIES, FROM THE YEAR 1772, WITH THE QUANTITY OF OIL, &c.

Year.	No. of Ships.	Tuns of Oil.		Year.	No. of Ships.	Tuns of Oil.		Year.	No. of Ships.	Tuns of Oil.		Year.	No. of Ships.	Tuns of Oil.		Year.	No. of Ships.	Tuns of Oil.	
1772	9	391		1782	3	217		1792	20	896	3 Clean	1802	34	2955		1812	49	6812	1 Captured
3	9	265	2 Clean	3	4	290		3	18	835	1 Clean	3	41	2262	2 Clean	3	54	3533	3 Clean
4	8	466	1 Lost	4	9	432		4	16	709	1 Lost	4	40	4018	3 Lost	4	57	7379	1 Lost
5	10	68	6 Clean and 2 Lost	5	16	722	1 Clean	5	14	1148	1 Clean	5	38	5165	1 Lost and 1 Captured	5	55	3607	3 Clean, 1 Lost
6	9	275	1 Lost	6	19	856	1 Clean and 1 Lost	6	17	1578	1 Captured	6	39	3524	3 Captured	6	55	5150	1 Clean
7	9	333		7	29	1132	1 Lost	7	21	1741	1 Lost	7	35	4350	2 Lost	7	57	4789	1 Clean, 1 Lost, 1 Broken up
8	8	171	2 Clean	8	34	958	3 Clean	8	23	2162		8	27	4556	3 Lost and 2 Captured	8	63	6219	1 Lost
9	3	142	1 Lost	9	27	854	2 Clean and 2 Lost	9	26	2244	2 Lost	9	26	4321	3 Lost	9	60	5077	3 Lost, 2 Broken up
1780	4	309		1790	23	832	2 Clean and 1 Lost	1800	22	1818		1810	34	5019		1820	60	7978	2 Lost, 1 Broken up
1	3	263		1	18	345	4 Clean and 3 Lost	1	24	2149	1 Lost	1	42	5398	1 Lost,	1	52	5888	1 Clean, 9 Lost

☞ The lost and captured Ships are not included in this number of Ships.

Opposite: Broadsheet of 1821 listing the 55 ships of Hull's whaling fleet.

Above: The Greenland Right whale, 60 feet long, from a drawing by William Scoresby. Other whales are shown in the two-page chart overleaf.

PIKE WHALE
Balaenoptera acutorostrata
33 feet

KILLER WHALE
Orcinus orca
30 feet

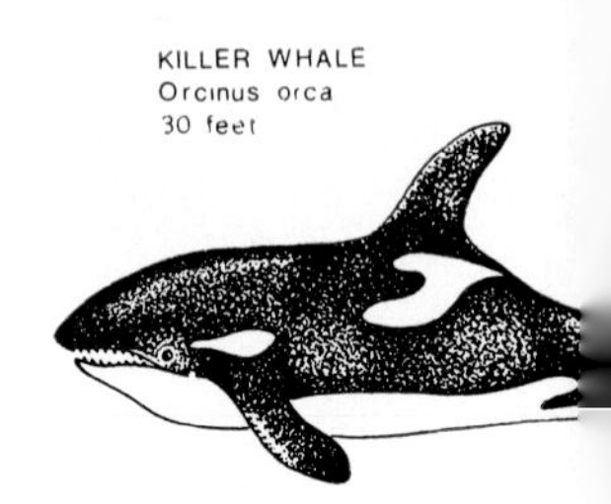

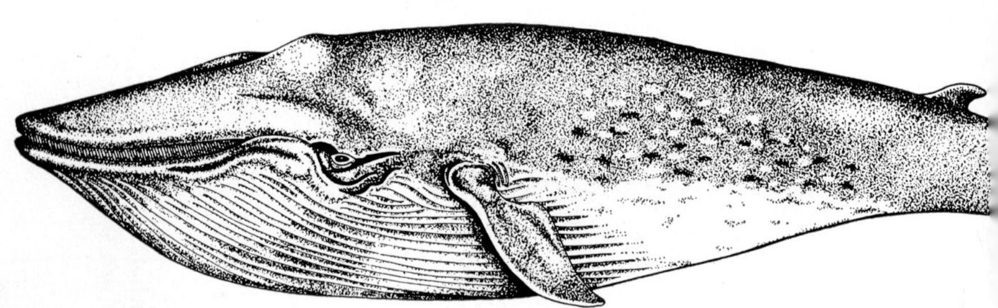

BLUE WHALE
Balaenoptera musculus
100 feet

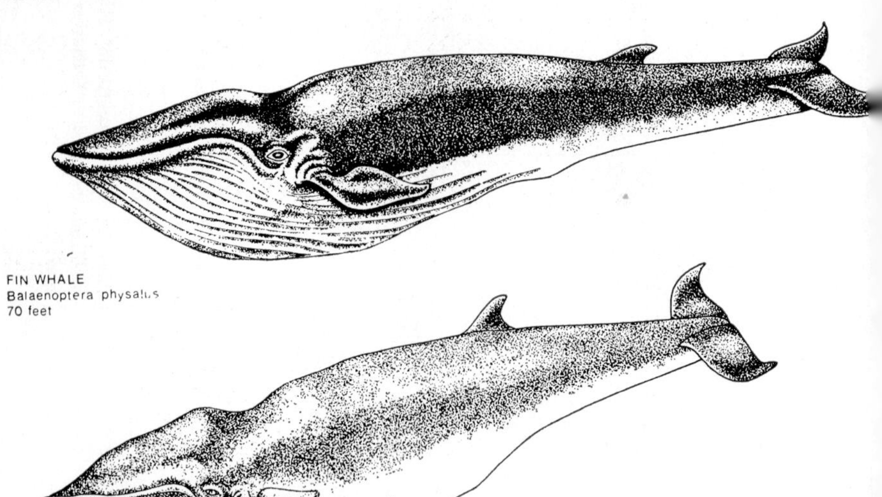

FIN WHALE
Balaenoptera physalus
70 feet

SEI WHALE
Balaenoptera borealis
55 feet

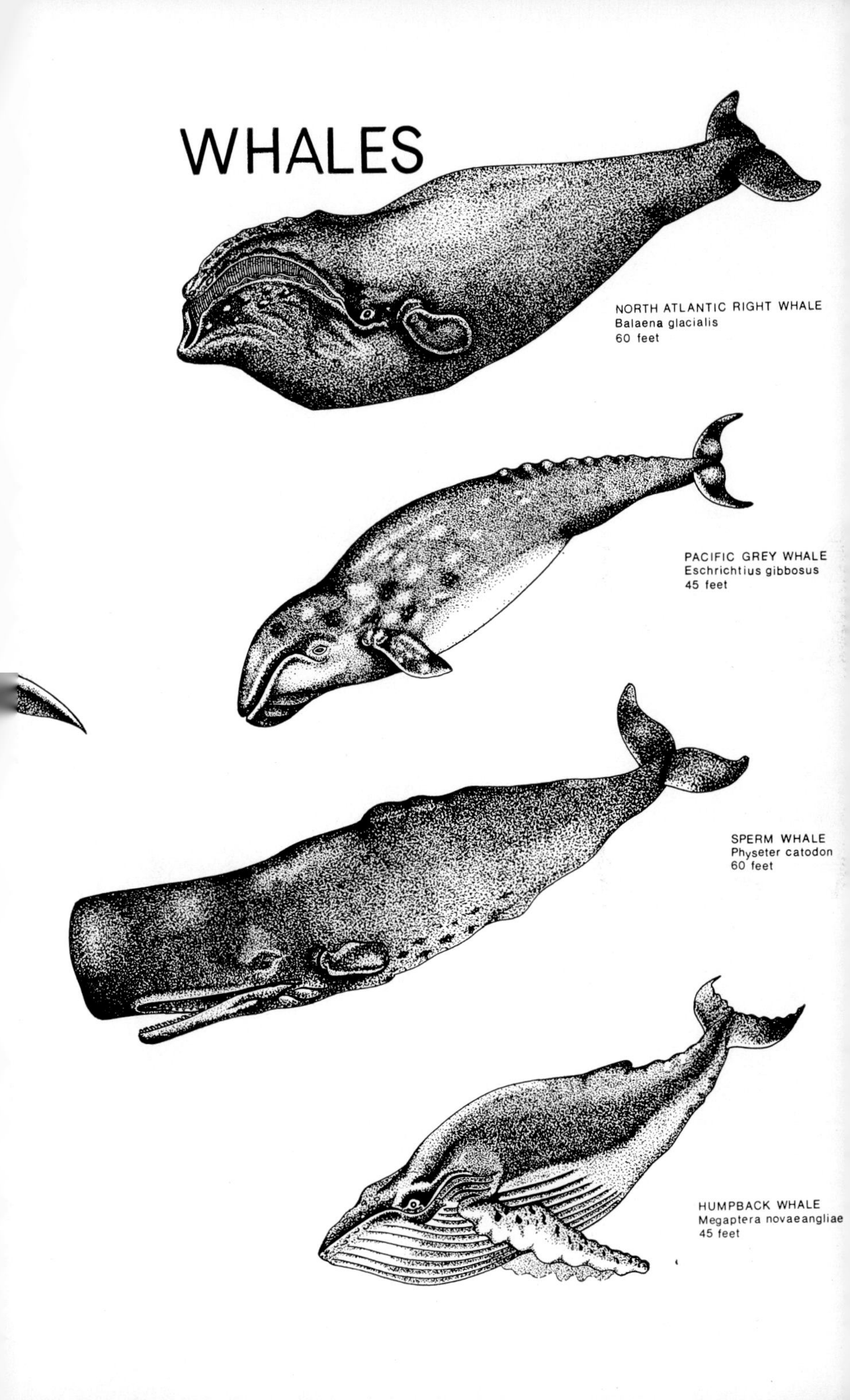
WHALES
NORTH ATLANTIC RIGHT WHALE
Balaena glacialis
60 feet
PACIFIC GREY WHALE
Eschrichtius gibbosus
45 feet
SPERM WHALE
Physeter catodon
60 feet
HUMPBACK WHALE
Megaptera novaeangliae
45 feet

Harpoon gun of about 1950 mounted on the bows of a whalecatcher. The explosive harpoon bursts within the body of the whale.

Whalebone scrapers at work: an illustration from Walker's 'Costume of Yorkshire,' 1814.

organs. Blood would spurt over men and boats, and the whale would boil the sea a frothy red in its death throes.

One reason that the Greenland whale was the 'Right' whale for the whalemen was that it floated when dead, unlike the longer, slimmer, and faster, fin whales. Yet though it floated, the whale boats might be several miles away from their ship, and first they might have to wind in a great bight of heavy line which the whale had taken out. After that, lines were fastened from each boat to the whale's flukes, and the towing began, perhaps 60 or 70 tons of whale acting pretty well like a dead weight, drifting steadily to leeward if the wind was brisk and the sea was choppy.

Back at the ship, the whale was stretched alongside and flensing began immediately, for if there was one whale there would be others, and also the vast carcase had begun to deteriorate and fill with gases the instant life had ceased. In 1818 a Hull captain, with 14 dead whales tied alongside, was unable to flense them because the intense cold had frozen the salt water ballast in all his remaining casks. Blubber was everywhere, but, he said, 'we lost nearly all these fish. They commenced to work and swell and burst alongside before we could flinch 'em, and was so spoiled we had to turn 'em adrift. A dead fish swells and bursts very quick, there's so much heat working in 'em, and when they bursts we can do nothing with 'em.'

The whale was flensed by men climbing over the vast, slippery, body, with spikes tied to their boots, and wielding huge, razor-sharp, knives. Blubber was peeled off the whale in strips, like skin off a banana, and on deck cut up and pressed into casks. The whalebone, which was actually the fibrous baleen fixed to the upper jaw and used by the whale as a sieve, in that it took in plankton-rich water and, by pressure from the giant tongue, expelled the water through the baleen whilst retaining the plankton, was cut away from the head, as perhaps would be the jawbones, and the carcase, known as the 'kreng,' would be cut adrift. It would sink, but before long rise again, becoming smelly and horribly bloated.

During the whole greasy and laborious operation birds would be around the ship: burgomasters, puffins, kittiwakes, and especially fulmar petrels, known as 'mallies' or 'mallymuks' by the whalemen. Sometimes they would aid in the killing of the whale, hovering over it and indicating its position to the boats when it lay just below the surface. Now, at the flensing, they squabbled in the water as thick as blades of grass on a lawn, getting under the feet of the flensers, and in their way. They gorged themselves, regurgitated, and returned for more.

The pinnacle of a season was reached if a ship became full, that is, with every tank and cask and hole and corner jammed tight with blubber. She could then set off back home, certain, if she made it safely, of a rapturous welcome from everyone. This was the dream, which happened often enough to seem possible always. The nightmare, however, happened perhaps more frequently, when a ship had scurried fitfully from one fishing ground to another, always too early, or too late, or blocked by ice, so that she returned a 'clean' ship, perhaps damaged, earning nothing for the owners, and signifying a harsh winter, and hardship through lack of food and clothing, for the families of the crew.

It was a brutally hard business, with never much in it for the deck-hands, even on a successful voyage. Captains were sometimes part-owners of the ship, and they, and their senior officers at least, often supplemented their wages by private trading with the Greenland Eskimos, exchanging knives and needles, clothes and trinkets, for narwhal horns, bear and fox skins, and garments made of reindeer hide.

The owners, who argued that as they took the greatest gamble so it should be they who reaped the highest rewards, also had a way to supplement their investment. Insurance premiums for the Arctic were high, yet, nevertheless, ship and probable cargo seem to have frequently been over-insured, so that a ship might show her greatest profit by being completely wrecked. Captain Gravill lost the *Chase,* crushed by ice, in Baffin Bay in 1860. An American-built ship, she had been bought for £3,500, strengthened, given new masts and engines of 140 horse power, and then insured for £16,500.

'. . . and I warrant you,' Captain Gravill said years afterwards, ''twas the best voyage the *Chase* ever made.'

Chapter Four

PRESS-GANGS

LIFE on a whaling ship was in some ways like an elaborate board game, where the dice decided whether a whale would be caught, or the ship would be trapped in ice; whether fair winds would blow, or the ship lose its bearings completely in days of continuous fog; whether a school of whales would present themselves, or whether, within sight of home sweet home, utter calamity would strike.

Naval press-gangs were active during most of the 18th century and well into the 19th, especially at the time of the wars with France, and their activities were condoned in law because an overwhelmingly strong navy was required to keep Britain unmolested, and her trade routes open, whilst conditions in the Service were so appallingly bad that there were never enough volunteers to fill the vacant places. A man was torn away perhaps never to return, unless he was maimed and of no further use to his country. Whalemen, who were considered to be amongst the finest of seamen, were prime targets for impressment.

'Protections,' in the form of official certificates exempting the individual from naval service, were issued by Customs to the whalemen, although generally only to the officers and perhaps to the harpooners. As Hull was one of the largest ports in the country, it received constant attention from the Navy, and the press-gangs operating in Hull were a constant menace, almost as unpopular with the landsfolk of the town as with the seamen.

The Customs were also interested in the returning whalers, for at various times whale products were liable to duty, and there was also, perhaps even in wartime, the constant risk that the whalers had been met by Dutch or French smugglers, and consequently would have casks of undutied gin and brandy, and packages of snuff and tobacco, hidden aboard to celebrate their return.

In the late August of 1761, the captain and crew of the *Berry,* returning from Greenland waters, were disgusted to be accosted in the Humber by both Customs and the Navy. The Customs boat, rowing across from the Revenue cutter, was in the lead and had almost reached the whaler, when, to the surprise and horror of the King's men, the naval boat, pulling lustily from H.M.S. *Mermaid,* fired at them, 'very smartly with musket balls,' as the Customs later reported to their Commissioners, in London. Shaken, but unharmed, they scrambled aboard to be quickly followed by an aggressive young naval lieutenant, who, in 'a Swearing disolute manner,' and with his boat's crew armed to the teeth with carbines, pistols, and cutlasses, gave as his reason for firing his belief that the Customs might be carrying protections against impressment to give to the Greenland men.

The *Berry* incident took place in the Humber, but in fact the returning whalers to Whitby and Hull were at risk virtually anywhere within British waters. In 1794, when the war with France was already three years old, and naval discipline was kept only by the authority of the lash, the Hull whaler *Sarah and Elizabeth,* returning from a season at Davis Straits, was accosted by the Navy off the Berwickshire coast.

Being a whaleman was hard enough, and often was only entered into because at least it was a job, whereas without one you stood a chance of starving. However, at least a man could cease being a whaleman after the first voyage, but this did not apply to the Navy, which kept you in for as long as it wanted and could hang you if you deserted and were later caught. The crew of the *Sarah and Elizabeth* went down into the 'tween decks and the hold, battening the hatches, and refusing to come out. Royal Marines boarded her, removed the hatch covers and fired down amongst the whalemen, killing Edward Bogg, the carpenter's mate, and seriously wounding two seamen. So incensed were the marines by the whalemen's defiance that one of them was prevented from throwing a grenade into the hold only by frantic insistence that if he did so, then the whole ship would blow up, killing him as well. Rather than be slaughtered like rabbits, the whalemen came up, and were dragged aboard the frigate, H.M.S. *Aurora,* to be shackled with leg and wrist irons.

In Hull, the coroner's jury brought in a unanimous verdict of wilful murder against the captain and part of the crew of the *Aurora,* although it seems to have been the practice of the Admiralty to shrug its shoulders at such civilian verdicts, and to ignore them.

The owners of the whaleship considered instituting legal proceedings against the *Aurora's* captain; and a few days later the *Hull Advertiser,* of September 13th, 1794, had a single paragraph in its close-printed columns, which announced: 'We Stop the prefs to say that fourteen of the harpooners, boat fteerers, and line coilers, together with the carpenter and fecond mate, who had been impreffed from the *Sarah and Elizabeth,* have been difcharged at the Nore.'

The activities of the press-gangs and the naval guard vessels became so bad around Hull that some crews insisted on being landed either on the Lincolnshire coast, from where they would make their own way home, or at least in the lower reaches of the River Humber. Four years after the *Aurora* incident, and with the war with France now seven years old, the returning whaler *Blenheim* succeeded in reversing the tables on the Navy.

This season, the single ship from Scarborough, the *Vigilant,* under Captain Clay, with ten whales, had arrived back in mid-July, and all the Hull and Whitby ships had arrived by the end of August. Those from Greenland waters had had a particularly good season, the *Volunteer,* under Captain Bedlington, arriving at Whitby in July with 265 butts of blubber from 19 whales, and the *Henrietta,* under Captain Kearsley, with 240 butts of blubber from ten whales and 155 seals.

Captain Scoresby, senior, this year in command of the London whaler *Rodney,* had caught 36 whales, which was thought to be the greatest number ever caught by a British vessel. The *Blenheim,* of Hull, under the command of Captain Mitchinson, with 18 whales, arrived in the Humber at the end of July.

As she sailed up with the tide from Spurn Point to Hull, and was within a short distance of Paull, she drew level with two anchored naval guard ships, the *Nonsuch,* and the *Redoubt.* The *Nonsuch* ordered her to stop, and when she did not, fired into her, though with light shot, for it was one thing to impress seamen, but an entirely more serious matter to destroy property.

The *Blenheim* continued to defy them, sailing steadily past, though whether under Captain Mitchinson's command it is not possible to say, for one report later had it that the crew locked him and the mate, neither of whom would have been impressed, within the captain's cabin, and took over the ship.

A third Navy ship, the sloop of war *Nautilus,* anchored nearer the town, also shot at the whaler, and launched a boat to join the two already chasing after the *Blenheim.*

The whaler managed to keep in front, although the three boats were being rowed lustily, and were closing. Here, at the entrance from the Humber into the River Hull, the helmsman swung the *Blenheim's* head too sharply, and she struck, and stuck, in the deep mud bounding the main channel. The three boats rowed around her, threatening destruction to the crew, loosing off muskets spasmodically, and attracting the attention of the whole town. All vantage points were soon black with spectators, on land, and on the ships in the harbour. There was no doubt that all, men, women and children, were completely on the side of the whalemen, cheering them on, jeering and howling at the Navy.

The *Blenheim* was armed with carriage guns, normally a protection against French privateers. They were now loaded with grape shot, which had a devastating shrapnel effect at close quarters. One was fired, not at, but deliberately near to, the naval boats.

The Navy drew cutlasses, rowed in, and tried to get aboard; the whalers grabbed their lances, and laid about with flensing knives. Blood flew in the pitched battle that followed, and limbs were broken. Before the Navy retired, startled by the desperate ferocity of the whalemen, one of the sailors had been killed, and two had been badly injured. In the respite, and helped by the surrounding shipping, the crew of the *Blenheim* escaped into the town.

On this occasion the *Blenheim* triumphed, but she was unable to do so a few years later when whaling in Davis Straits. The war was still going on, and a French frigate took off the *Blenheim's* crew, applied the torch, and moved further away as the grease-and-oil-encrusted veteran burned ferociously, giving off billowing clouds of yellow-black smoke.

Chapter Five

THE GREENLAND MEN

THE sea has always been a hard taskmaster, and, when combined with ice and fog, and a 'rabbit' philosophy, which accepted that the lower orders would always supply cheap and expendable labour, the odds were stacked against the whalemen and their ships. The Northern Fishery must have been at the top of the league for danger and discomfort, challenged only by coal mining.

There were about 50 men and boys to the average whaler, and a comparison over the years between the ships which set out so bravely, and those which returned, shows a tremendous mortality rate. Nor was the danger confined to the high latitudes; ships disappeared, as did the *Royalist,* of Hull, in 1814, with 54 men and boys, leaving 30 wives now as widows, and 100 children fatherless. The following year the *Clapham,* of Hull, blew up in the North Atlantic, when fire spread to the powder room. Nearer home, there was always the Yorkshire coast to contend with, a graveyard for sailing ships. In 1826 the returning *Esk* was wrecked off Saltburn; and Hull's whaling days ceased in 1869, with the wrecking of the *Diana* at the mouth of the Humber.

Some men made whaling their way of life, sailing in February or March and returning in autumn or early winter, year after year, working their way, if they were capable and ambitious, from seaman to harpooner to officer, and even through to captain, and part-owner. Many went years without seeing either an English spring or summer —Captain William Barron of Hull, wrote in his autobiography that, for 17 years, he saw neither blossom nor fruit on the trees, nor flowers, nor ripened corn and harvesting. Many Shetlandmen also returned to the whaling year after year, often to the same ship and captain, although they were invariably signed on as seamen, and rarely rose any higher. One of the predominant reasons—perhaps the main one —sprang from the perpetually deprived state of the Islands. However bad conditions at the whaling might be, they were better than in Shetland, dominated by avaricious lairds, with their ruthless factors; sombre, hell-fire ministers; morbid superstition; and endemic poverty.

To qualify for a government bounty, each ship had to take a surgeon, and most Yorkshire vessels continued to do so even when the bounty was withdrawn in 1824. Initially the appointment called for a surgeon/cook, which illustrates both the lack of standing that the medical profession had in the 18th century, and the owners' short-sightedness in not having a specialist cook, often the most important man aboard, and someone who could not only ruin a voyage, but be instrumental in saving life or hastening death when a ship became trapped in ice, and on short rations. William Scoresby, senior, has been accredited with being the first captain to employ medical students, on low wages, but eager and willing to go for adventure, and to obtain practical experience. Such doctors quickly attained a position of importance, dining with the captain, a source of intellectual stimulation to those who wanted it, and, to the sick, a purveyor of potions, and a sawer-off of gangrened limbs.

The seamen lived in the fo'c'sle, which was as dark and strong-smelling as a fox's lair. The only light was that provided by a whale-oil lamp, the fumes of which could not escape, so that carbon, in the form of soot, was apt to quickly cover walls and ceiling. The only ventilation was through the companionway door, which was kept shut most of the time to try and retain the heat of the single iron stove, which at least had a flue pipe passing through the deck to the open air. Bunks, lining the planking of the hull, were wide enough for two or three men, which saved space and conserved heat. Each man had a kit-bag, or a sea-chest which acted as locker, seat, and table. When there was snow outside, slush was tramped inside. And in really cold weather, of which there was an abundance, frost, and, in the winter, ice, coated bunks and walls, melting into a miserable wetness when the temperature rose.

Nevertheless, the dirt and the discomfort were accepted as part of the game, and were not looked upon as anything particularly unusual, except sometimes by the doctors, who came from middle-class families and found the conditions repugnant. At the beginning of the voyage, with the ship scrubbed and tarred and painted, there was nothing to suggest the floating slum it would become before it finished whaling, stinking of slaughtered seals and whales, with a crew shaggy and unwashed, and with both ship and men impregnated with claggy, slippery, grease. The Arctic land and sea and ice were almost clinically clean through the intense cold and lack of pollution; and into this the whalers sailed around like floating abominations, staining ice and sea red with blood, and leaving carcases of whale and seal and walrus, as they strove to stuff ever more decomposing blubber into the oil casks.

None of them seem to have expressed any distaste or hesitation at the killing of whales, although with young seals it was different. Adult seals gave five or six gallons of transparent oil, and a skin used in handbags, or by shoemakers. The babies, which have large, soft, eyes, and cry almost like a child when they are attacked, gave their fur, for ladies of fashion.

A ship could slaughter many thousands of seals, and as the weather was quite often bad, and the seals did not stay if they could possibly escape, an attack on a body of seals was a frantic, urgent, business, with the whole crew working extremely long hours without food or drink, crushing skulls with a pick, and ripping off the skins. The work was hard, for a pack of seals represented an open-air slaughter house, within which all had to be killed. Seals are full-blooded creatures, and, although men washed if they could beg hot water from the galley, hair and beards and clothes, impregnated with oil and blood, tended to stay that way.

With whales, the actual killing was almost as much a test of stamina and strength as it was of skill and experience, especially as the only motive power in the six or seven boats to a ship was provided by the oarsmen. After the kill, there might be a row of several hours to tow the whale back to the ship, to be followed by immediate flensing and then the filling of the casks.

It was generally a trade for young men, at least at the crew level, men fit and hard enough to accept the physical discomfort, and to revel in the risks and the gamble. Some signed on for adventure, others for notoriety, for, during whaling's hey-day at least, a Greenlandman was more than a mere seaman.

Many Hull seamen served on the Scandinavian and Russian trades, and knew about sub-zero temperatures and ice, whilst Whitby seamen excelled in most branches of trade at sea, yet a coasting voyage to Tyne or Thames, or a deep-sea one to Riga or Cadiz or Quebec, had none of the peculiar glamour of the Northern Fishery. The former type of voyage had a clearly defined cargo to be taken and to be brought back, on a far from generous wage; but the latter had no fixed destination, and a possibility of returning wealthy, or comparatively so. On a routine cargo trip, dangerous and arduous though it might well be, the seaman tended to be a nobody, little more than a pawn of masters and owners and elements, whereas in the whaling, working generally on a bonus system, he was a definite part of a single unit.

All benefited, or lost, from a good or a bad voyage, and all shared common dangers, from senior captain to junior apprentice. The constant danger gave a swagger to the whaleman's walk. It was expected that a cargo ship would return, and it was a calamity when it did not, but at the whaling it was expected that at least one ship would sink, or disappear, each season—almost one in every five did so in 1830, out of a combined English-Scottish total of 90 ships—and so an element of bravado entered into the business, as if the rest were enlisted men, whilst the Greenlanders were picked volunteers.

For whatever reason, whaling attracted men from different parts of the country, although most lived locally. The crew list of the *Ellison,* for example, out of Hull in 1797, shows the majority of the men as being from Hull, but also men from Whitby, Shields, Salisbury, Pontefract, Howden, and York. The *Perseverance,* out of Hull in 1809, had individuals from Somercoates, Bridlington, Lynn, Durham, Shields, Newcastle, York, and Cottingham.

The spirit created by shared dangers, shared incentives, and comradeship, could be exploited by an astute captain—and records suggest that most of them were—for the captain's influence was paramount. If he was respected, strict on discipline, perhaps, but fair and honest, then it was a good ship to be in. Crowds of relatives and well-wishers would give the ship a noisy farewell, with a garland of ribbons, provided by wives and sweethearts, tied to a stay near the mainmast head. Crossing the Arctic Circle, Neptune and his attendants would visit the 'green men,' fastened to a chair whilst they were lathered with a mixture of fat and flour and then shaved with a huge piece of rusty hoop iron. Amongst the ice, games of follow-my-leader would be played if the ship was waiting for a wind, up the masts and out on the yard-arms, several score feet above the deck, then down by the backstays, to leap-frog the capstans, and even up onto the bulwarks and finally ending by diving into the icy sea.

Most captains carried a few casks of rum, and no doubt this was an added attraction to some men, as were the calls at Lerwick, Shetland. If a vessel was to go to Greenland sealing first, then it called at Shetland to make up the crew and take on provisions, returning perhaps a month later to offload its seals and stock up for the whaling. At the end of the season there would be a third call, to discharge the Shetlandmen. Three calls at a town regarded for a long time as being the wildest in Britain, with a surfeit of *usquebaugh*—immature whisky—women, and song, were a bonus for many of the whalemen, and they involved themselves with abandon. In 1826, Captain Munroe, of the *Cumbrian,* of Hull, recorded in his log that some of his crew had gone to the funeral of a seaman off the *Lord Wellington,* also of Hull, who had died 'in consequence of taking spirits in excess on shore,' and, a day or two later, Munroe's own sailing was delayed, 'due to part of the crew being absent.'

One might have expected that the difficulties and dangers of a whaleman's life would have perhaps been compensated for by good food and generous wages, but this was not so. Some Yorkshire owners were positively frugal when it came to provisioning their ships, and, whilst these were balanced to some degree by a few at the other end of the scale, the majority gave only enough to keep their men alive and operating for six to eight months. This was sufficient if the ship returned as planned, but if she was beset by the ice and her return was delayed for several months, then deaths were virtually certain.

The staple diet was one of salt pork or beef, and ship's biscuit. Flour, for bread and the inevitable plum pudding, was also taken in quantity, and on some ships fresh sides of meat hung from the mast heads on the outset of the journey, covered with hessian against the depredations of the gulls. Rice, oatmeal, cheeses, and potatoes, were also taken, the last-named being probably the most popular item of food within the small stocks sometimes taken by individuals for their own use. The potato had the advantage of containing a small amount of ascorbic acid, or vitamin C, and helped in combating scurvy. The staple drinks were tea and coffee, especially the former, taken black, and heavily laced with sugar.

Tobacco—it was an age of pipes, and, for the officers, perhaps cigars—was an indispensable item, associated with relaxation, and having the priceless quality of dulling an active appetite. On the *Diana,* of Hull, which was trapped by the ice of Baffin Bay in 1866 and had food to last on full rations only until about the end of October, the doctor noted, January the 11th, 1867, that he had now finished his portion of the cavendish taken from the dead captain's stock, and the men, who had used up their own some time earlier, and had experimented with tea-leaves and bits of rope, 'are now becoming increasingly fractious towards each other.' The previous day he had written, after finding the first signs of scurvy, 'I am the doctor of this ship, the one to whom they will look for life and health, but look, alas, in vain. God help me. What can I do but trust in His mercy and pity and power to save?'

Under such circumstances as the *Diana* was experiencing, the owners in Hull or Whitby might pay half wages to the relatives of the men, until they either returned, or were adjudged to be dead. This kept aged parents or wives and children from actual starvation, or the workhouse, but did little more, for the system worked on a low basic wage, supplemented, if the voyage was successful, by bonus payments for the tons of oil and whalebone obtained.

The wage structure was geared to the owners' benefit, and in extreme cases, as with harpooners, who were paid purely on results, a bad voyage and an empty ship generally meant that each harpooner spent six months of hard work, and yet returned in debt to the company for the tobacco and drink and clothes which he had bought on the voyage.

There were no standard wages, and quoting a wage, for example, which was paid to a boat-steerer in 1850, on the old *Truelove,* of Hull, of 30s a month plus 2s 6d per ton of oil and 5s per ton of whalebone, gives only an approximate idea in the absence of comparative prices. On the same ship, Francis Gifford, a Shetlander, from Bressay, received £5 4s 6d wages for six months, 16 days, plus oil money of 16s for the three whales caught. His expenditure for the same period was £5 16s, for clothes, food, and tobacco bought on the ship, and he received a final payment in November of 3s 6d. However, along with the rum, 3d; brandy, 1s; twine, 1d; coffee, 10d, and so on that he had needed, he seems to have finished the voyage with a coat, 5s; trousers, 3s; drawers, 2s 6d; jacket, 13s; stockings, 1s 6d; cap, 1s 8d; two shirts, 2s 3d; and a skirt for his mother, 2s 6d, which she purchased at the Lerwick chandlers, in her son's absence.

Footnote: Cavendish was tobacco softened and pressed into solid cakes.

Chapter Six

THE WRECKING OF THE 'ESK'

BUILT in Whitby in 1813, the *Esk* was one of the town's most successful ships, though not one of the luckiest. She was a strong, sturdy, vessel, which had brought home good catches several times from both Greenland and Davis Straits. For 13 years she had left the North Riding in late winter to face the foulest of Atlantic weather, and to undergo the trials of blizzards, black ice, bergs, uncharted waters, and moving ice fields. In her third year, under Scoresby, junior, she had managed to get back from the Greenland Sea with part of her keel torn off. Now, on Tuesday, September 5th, 1826, off South Shields and with the blubber of four large whales in her casks, she was within a few hours of berthing in Whitby harbour.

It was at this stage that, for some unrecorded reason, the second mate of the ship had a violent disagreement with the Master, presumably a culmination of an existing friction between the two, for it now resulted in him insisting on leaving the *Esk* at once, which he did by a Shields' pilot coble. Perhaps, as whaleship and sailing coble parted, he had cooled down enough to regret having prejudiced his future, yet never was a temper better lost, for the impulsive decision to leave saved his life.

The *Esk* continued home, having to tack against a growing southerly breeze, and making slow progress, being built more for strength than for speed, and, as with all square-riggers, unable to stand close to the wind. When past Hartlepool, in late afternoon, with the day cool and hazy, she took in some sail and moved closer inshore, presumably to get the full benefit of the part-run, south-flowing, flood tide. Slowly she sailed past the mouth of the Tees and picked up the Yorkshire shore line, each familiar landmark seeming to draw them closer to the wives and sweethearts, parents and children, whom they had not seen for half a year. Even the circling gulls seemed to have a homely look about them. The Shetland men had been discharged at Lerwick, and all the 27 remaining men and boys must have felt that special sort of anticipation reserved for returning travellers from distant, dangerous, parts. Yet it was now that a fierce and sudden gale blew up, out of the empty sea to the east, pushing the *Esk* towards the shore, and shallow water.

The gale increased ferociously, with a frightening rapidity, tearing and ripping sails before they could be reefed. The crew fought, frantically, attempting to bring the ship into the wind, whilst the screeching gale, darkening the sky and bringing with it squalls of stinging rain, attempted to pluck them from the yards and throw them into the spume-torn sea. The ship had weathered many such storms, yet always in open sea, never this close to a lee shore.

The Whitby men were now literally fighting for their lives, for they had been too close in for this kind of weather, and now were unable to force the ship out into deeper water. Every last bit of hard-won experience was involved, every last ounce of energy was used to force the tiller around and brace the sails, yet, night was closing in as the great mainmast topsail split, tore apart, and then ripped into pieces which streamed horizontally in the howling wind. The *Esk* immediately lost any steadiness and manoeuvreability, swung broadside to the gale, and, despite anything that the fighting, despairing men could do, drifted inexorably towards the shore.

By half past ten, with the tide almost at low water, she struck bottom, an impossible stretch of crashing breakers and murderous undertow between ship and shore. Most of the crew gathered in the 'tweendecks; several prayed.

Seeing lights on shore, they fired the guns, and burned a blue distress light, althought it must have been increasingly clear that there was little they could hope for, with wind and sea becoming even wilder. The crashing waves, exploding against the hull of the *Esk* and climbing her masts, were now tearing away, or smashing, the whale boats.

The ship, although a tough old warrior, now began to open as the sea, with unbridled violence, continually lifted her up and then slammed her down against the sand.

As the hold filled with water, so the casks were lifted and took on the surge of the sea, battering the ship from within. All hands were now on the open quarterdeck, the least precarious part of the disintegrating ship, even though frequently washed by the sea. Incredibly, they were still there in the cold and watery dawn, although the *Esk* was by now so weakened that every crashing sea shook her as if she was palsied. The mainmast had collapsed, but, still partly roped, was swinging violently about the deck, which was half submerged. In the half light, through the mist of spume and spray, the whalemen could see a lifeboat trying to get off the beach to them—but it could not, due to the facing wind and the pounding waves.

At fifteen minutes past five, with the sea a liquid inferno, the ship disintegrated completely.

Twenty-four of the 27 crew died. Of the three washed up, somehow escaping the fierce undertow, William Pearson actually managed to stagger ashore by himself; William Leach, vomiting and helpless, was pulled out to recover fairly quickly; and Matthew Boyes, who had been crippled on the whaling grounds by a cut from a flensing knife in his leg, and who was washed ashore unconscious amidst a tangle of wreckage, was pulled out as a dead man, slowly to recover over a period of hours.

By mid-morning the *Esk* was merely a litter of wreckage scattered along the tide line; spars, casks, unidentified pieces of broken timber. Later, as wind and sea died down, there were the bodies.

The year was a tragic one, for this was the second devastating blow Whitby had suffered. Earlier, in April, the *Lively* had been crushed by the ice of the Greenland Sea, leaving no survivors.

Altogether, 65 local men and boys had been drowned. Twenty-six wives were now widows. Eighty-one children were now fatherless.

Extracts from the log of the 'Laurel' —

(See illustrations on pages 42 and 43.)

May 31st, 1827:

'Forepart light Breezes from NE and clear Weather. At 4 p.m. got done Making off this fish filling 37 Butts, cleared the Decks and set the Watch. Dodging at the Edge of the Pack. At 6 p.m. saw a Whale, called all Hands and sent 6 Boats away. At ¼ past 6 p.m. Isaac Lindsay Struck her, run out 3 lines. At 7 p.m. Killed her upon 4 Harpoons. At ½ past 7 p.m. got her alongside, found a Harpoon belonging to the Princess Charlotte of Dundee and 8 lines. Made the Lines fast to the Ship, at 9 p.m. began to Flinch, at ½ past 2 a.m. got done. Length of bone 11 ft 1 inch. Cleared the Decks and set the Watch to Heave the Lines in. At 8 a.m. fresh Breezes from NNW and clear Weather. Reefd the Top Sails etc. Breezes and Weather to the End. No fish seen.'

Monday, July 28th, 1928:

'Forepart calm, dense flying haze, towing as above, *Swan* and *Rhambler* etc in company. Middle part gentle breezes and dense haze, plying towards land in pursuit of the boats. At 12 midnight met 2 towing the last stated fish, and a fish which Alex Markham had struck. At 1 a.m. made fast to a floe, the 2 boats crews going to rest. At 4 a.m. all hands called, one boat's crew preparing for flinching, the other returned to the 4 boats on watch. Latter part calm and foggy, at 9 a.m. the 5 boats returned towing a fish, which Ed. Hoodless had struck, having been on watch 48 hours. At 10 all hands sent to rest, so ends this day.'

Chapter Seven

SHIPS AND ICE

THE longest lived of all whaleships out of England and Scotland was the 300 ton, three-masted barque, *Truelove,* out of Hull, which first sailed to the whaling in 1784, and finished as Hull's second last ship in 1868.

She was built in Philadelphia, Pennsylvania, in 1764, and seems to have come to Hull as a prize taken during the American War of Independence.

During the whole of her long life there was an increasing rivalry between the British merchants and their upstart, wide-awake, American cousins, and certainly the *Truelove* was a fine advertisement for Yankee shipbuilding. Her strength and shape ensured that her whaling owners could confidently expect her back at the end of each season, which was an expectation not held for many other whalers. Few accidents were caused by impact with the ice, for the whaleships were rigged to be instantly manoeuvreable, but many were 'nipped,' or crushed, and it was here that the *Truelove* triumphed. Her build was such that she sloped inwards sharply at the waterline, in a manner which encouraged the closing ice to lift her, so that she rose above it, instead of attempting to resist the increasing pressure.

Year after year, interspersed with a little general cargo carrying to the Mediterranean, or to the Baltic, the *Truelove* returned to the whaling, a sound, steady, vessel, which generally made a good average catch, and which seemed to be able to take anything the Arctic could throw at her. In 1856 she was trapped in the ice twice, the first time when sealing in Greenland waters, when a change of wind and a plumetting drop in temperature resulted in 50 ships from several nations becoming beset—Captain Wells wrote, 'I cannot, in my anxiety, avoid wishing that the Devil had all this ice that now confines me; for I am quite sure, if transferred within the pale of his dominions, it would afford him a plentiful supply of hot water for some time to come'—and the second occasion was when whaling in Baffin Bay, this time along with two Scottish ships. All three were temporarily abandoned, and one, a Dundee vessel, under the command of Captain Deuchars, was completely wrecked, which resulted in the crew getting at the rum, and quickly becoming fighting drunk. Back home, there was an acrimonious correspondence in the press between Captain Wells and Captain Deuchars as to what had actually happened, the Scot asserting that he had had one of the best behaved crews in Baffin Bay, and the Englishman doubting whether the Scot's eyesight and memory were as good as they should be.

Despite her advantageous shape, the *Truelove* was typical of most ships in the trade in that she was not specifically designed for it, and until the purpose-built Dundee whalers, which were superior in all ways to the English ships, began to be launched in the 1860s, most whalers were converted from a miscellany of general sailing vessels, cargo and passenger, and in fact Whitby's veteran was one of these, the *Volunteer,* which made 54 successive trips. Fishburn and Broderick, whose shipyard was where the British Rail goods yard now is in Whitby, built the *Resolution* in 1803, and the *Esk* in 1813, and the *Baffin* was built for William Scoresby, junior, in Liverpool in 1820, but these were the exceptions. All three ships made consistently good catches—in 1814 the *Resolution,* under Captain Kearsley, brought back a record amount of blubber, which produced 230 tons of oil—but the sea claimed them all. The *Resolution* was sold to Peterhead owners in 1829 and lost in the Arctic the following year; the *Esk* was wrecked off Saltburn when returning in 1826; and the *Baffin* was sold to Leith owners in 1825, and lost at the whaling in 1830.

The *Sheffield Daily Telegraph,* of April the 13th, 1867, wrote that the typical whaleship had bows that 'bulged beligerently, like the swollen knuckles of a prizefighter,' and, certainly, one of the requirements for a whaler was great strength. Each converted ship was doubled or trebled, which involved fixing one or two extra exterior layers of oak planking around the bows, tapering to midships. Half inch iron plates were often fixed over these planks, and at other vulnerable points. Within, the hull was strengthened by an arrangement of 12-inch square timbers, 25′–0″ long, known as 'ice-beams,' which, together with secondary woodwork, not only strengthened the ship against impact, but also dissipated the force of a blow over a wide area.

The ships were usually three-masted and square-rigged, between 300 and 400 tons registered weight. Until the mid-19th century all vessels were purely sail, but, through the growing use of marine steam engines, the *Diana,* of Hull, after a successful first season, was fitted with a single screw in 1856.

She had been built in Bremen, of oak, in 1840, and used as an emigrant ship taking the poor of Europe to a new life in America. First registered in Hull in 1856, she was 355 tons, 117′–0″ long, 29′–0″ broad, and with 17′–0″ depth of hold. Converted by strengthening, she was an average whaling ship; but converted to steam she became the first of the new generation of auxiliary screw-steamers, with, according to the *Shipping Register,* two engines which weighed 63 tons, but gave her only 40 horse power. In theory, she was now going to be independent of the wind, but in practice, laden down with the engine weight and the great amount of coal she had to carry, she could not overcome any wind beyond a light breeze.

Generally, speed of passage was not essential, although one of the secrets of Scoresby senior, probably the most successful whaling captain in the country, was that he trimmed sails and ship so finely that on the whaling grounds he could invariably leave the others behind, and get amongst the whales before they arrived.

Manoeuvreability, known as 'handiness' to the whalers, was the prime requirement, for the whaling grounds became increasingly dangerous as the diminishing numbers of whales retreated further and further into the pack ice, which one moment might appear to be as static as land, and the next, without any warning, was in movement. Channels closed, holes filled, ice rafted, and ships which could not get out of the way were nipped. Nor was it only within the ice fields that danger constantly lurked, for pushing through south-flowing currents thick with bergs was a hair-raising experience. Sleet and snow, fog and high winds, made conditions bad enough by day, and heightened the risk at night. If the lookout saw a berg, as a dim luminescence looming out of the blackness, it was absolutely imperative that the ship twisted away like a frightened hare. The disaster that happened otherwise is illustrated by the story of the whaler *Shannon*, of Hull, in the April of 1832.

In a full gale, the sleet blinding down with almost flesh-cutting velocity, she had entered into the East Greenland current, which sweeps southerly past Cape Farewell, at the southernmost tip of the vast island, carrying in it a thick jumble of ice. At 3 a.m. the *Shannon* hit a berg. The impact broke the fore and main masts, the ship turned broadside to the ugly waves, and almost immediately began to break up. By the time the survivors had rigged a sort of canvas shelter on the tilting fo'c'sle, the after part of the ship was almost submerged, the sea sweeping over it. Sixteen men and three boys had been washed away. The survivors lashed themselves to the woodwork, and, when dawn broke, managed to salvage a cask of flour and some salt beef. There was no fresh water. Heavy seas were now breaking over most of the ship, which, though kept afloat by the buoyancy of the casks in the hold, was floating with only the fo'c'sle fully clear of the water.

By the fifth day, two men had died, and several of the survivors were delirious. Others pleaded with the doctor that he should bleed them, so that they might appease their thirst. He did so for 18 of them, some drinking the blood as it oozed from their veins, others mixing it with flour. Two days later, they were picked up by two Danish cargo ships. Frostbitten and gangrenous limbs had to be amputated, and seven more died. The Danes, who did all they could for the survivors, took them back to Copenhagen.

The *Shannon* was catastrophically unlucky, although mishaps before the whaling grounds were reached were not uncommon; the *Phoenix*, for example, was wrecked outside Whitby harbour on her outward trip in 1837; the *Fame*, out of Hull in 1823 with Scoresby senior as captain, was destroyed by fire at Kirkwall; and the *Corkscrew*, a large iron sail-steamer out of Hull in 1859, lost her screw north of Iceland, and had to return after catching just five seals. Nevertheless, most vessels did manage to reach the whaling and sealing grounds intact, and could begin their search for oil and skins.

Initially, at the beginning of the 17th century, the whales of Spitsbergen were numerous and easy to slaughter, and stinking whaling stations began to litter the shores of Spitsbergen, with the Dutch calling their foremost one Blubbertown. By the second half of the 18th century, these waters had been almost fished out, and the whalers

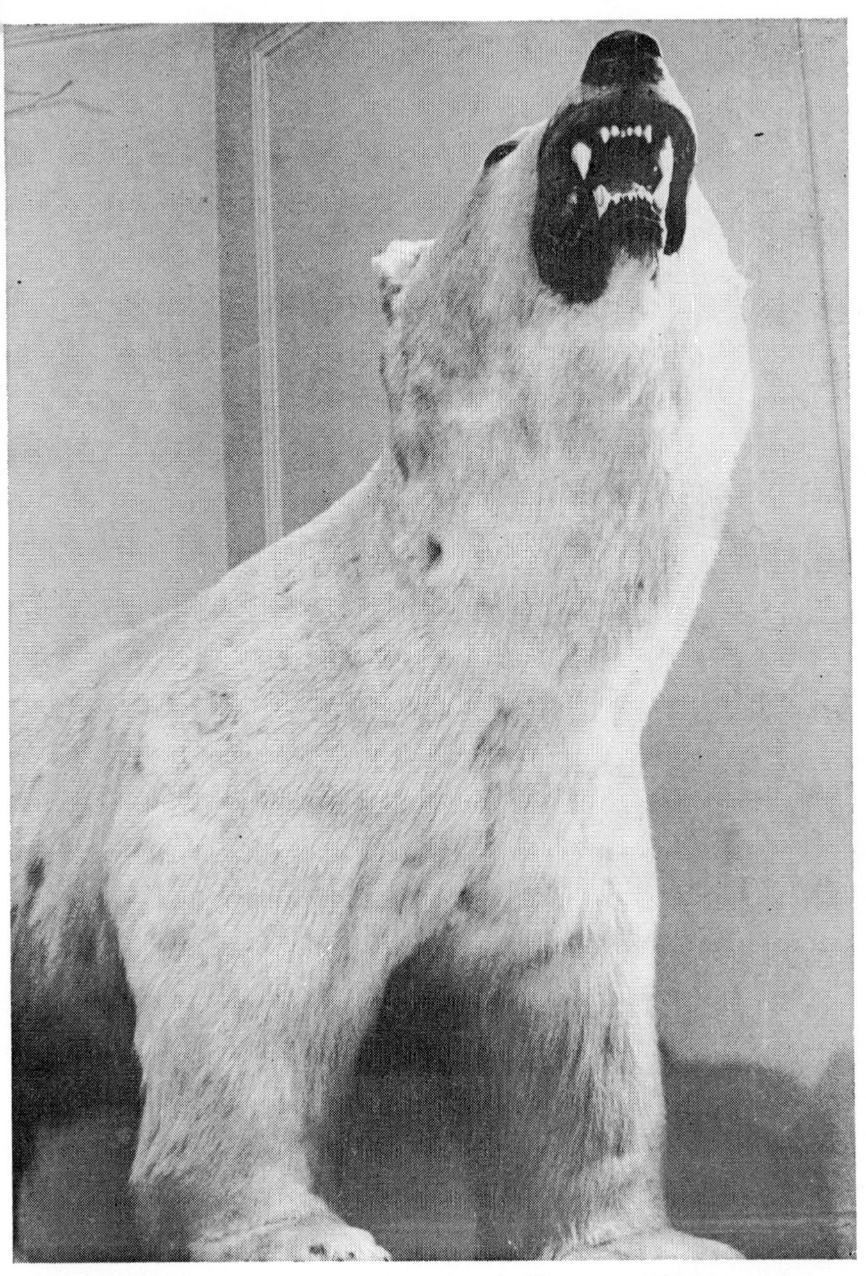

Polar bears — the largest of all carnivores. They were killed by whalemen for the fur, the cubs being brought back to England and sold to fairs and zoos.

Remarks at the South West 1827

Wedensday May 30th

Fore part fresh Breezes from E.N.E. and clear Weather plying along the Pack at 7 P.M. saw a Whale sent 2 Boats away and at 9 P.M. took them on Board without success Middle part moderate breezes from E.N.E. and Do Weather Watch employed turning Blubber out of the Main Hold ready to make off. at 4 A.M. called all Hands to make off Latter part light breezes from N.E. and clear Weather all Hands employed making off. (Duncombe of Hull) and several more sail in sight

Lat Obd. 61.42

No 5
ft in
11.1

Thursday May 31st

Fore part light Breezes from N.E. and clear Weather at 4 P.M. got done Making off the fish filling 37 Butts, cleared the Decks and set the Watch Dodging at the Edge of the Pack at 6 P.M. saw a Whale called all Hands and sent 6 Boats away at ¼ past 6 P.M. Isaac Lindsay Struck struck her run out 3 Lines at 7 P.M. Killed her up 4 Harpoons at ½ past 7 P.M. got her alongside found a Harpoon belonging to the Princess Charlotte of Dundee and 8 Lines made the Lines fast to the Ship, at 9 P.M. began to Flinch at ½ past 2 A.M. got done length of bone 11 ft 1 in. cleared the Decks and set the Watch to Heave the Lines in at 8 A.M. fresh Breezes from N.N.W. and clear Weather Reefd the Top Sails Do. Breezes and Weather to the End no fish seen

Lat Obd. 61.34 N

Extract from the log of the 'Laurel' in May 1827. The entries for May 31st and July 28th, 1828 (opposite) are reproduced on page 37.

Journal on board the Laurel in D…

Dates	Winds	
1828 Saturday July 26 —	Calm W.S.W.	Fore part, calm, 2 boats on watch, all other hands 'making off'; at 7 P.M. got done, 41 Butts made off; being the blubber of the last mentioned fish, clean'd the decks & set the watch. Middle & latter parts Calms & slight airs, dodging among loose Ice Swan, Abram & several Scotch ships in Co— at 8 A.M. all hands called, 5 boats on watch Cooper & others setting up shakes to the end
Sunday July 27 — L.C.	Calm — F.10-·—	Fore & middle parts, calms & light airs from the Land at 1 P.M. 6 boats on watch, among the Ice toward shore, at $8\frac{1}{4}$ P.M. Leonard Conyers struck a fish, at $9\frac{1}{2}$ killed her, & got her in tow at $10\frac{1}{4}$ P.M. 2 boats came along side, 4 remaining on watch at the Land floe. Latter part, calm, & hazy at 5 A.M. a boat came to ship for provisions & returned with the 2 boats along side to the Land floe, jolly boat towing ahead toward the Land to the end. Coutts Inlet, bearing by Compass N.W. by W. dist. 25 miles.
Monday July 28 — A.M. — E.H.	Calm Variable — F. 9"9 In. — F.10-3-	Fore part, calm, dense flying haze, towing as above, Swan & Rhambler &c. in Co. Middle part gentle breezes, & dense haze, plying towards Land in pursuit of the boats, at 12 midnight met 2 towing the last stated fish, & a fish which Alex Markham had struck, at 1 A.M. made fast to a floe, the 2 boats crews going to rest, at 4 A.M. all hands called, one boats crew preparing for flinching, the other returned to the 4 boats on watch. Latter part, calm & foggy at 9 A.M. the 5 boats returned towing a fish, which Ed Hoodless had struck, having been on watch 48 hours, at 10 all hands sent to rest, so ends this day

A July 1828 extract from the log of the 'Laurel.' This shows three whales killed, denoted by the inked-in flukes, with measurements of the baleen or whalebone.

The 'Munificence,' from a painting of about 1805 by Robert Willoughby.

The 'William Lee' in the Arctic. This and the painting on page 46 are by John Ward (1798 - 1849).

'Swan' and 'Isabella,' a painting executed to celebrate the rescue of the 'Swan' after a winter in the Arctic ice **(see pages 49 - 50).**

'Diana,' last of the Hull whalers, off Cape Searls (see chapter eight).

Lecture-Hall, Goodramgate, York.

THE TWO ESQUIMAUX OR YACKS,

Male and Female, brought home by Captain Parker, of the Ship *Truelove*, of Hull, from Nyatlick, in Cumberland Straits, on the West side of Davis' Straits,

WILL BE EXHIBITED

On Thursday and Friday, March 9th & 10th,

In the Lecture-Hall, York,

For Two Days only, previous to their return to their Native Country on the 20th Instant.

This interesting married couple, MEMIADLUK and UCKALUK, (whose respective ages are 17 and 15,) are the only inhabitants ever brought to England from the Western Coast. They have been visited by upwards of 12,000 persons in Hull, Manchester, Beverley, Driffield, &c.

THEY WILL APPEAR

IN THEIR NATIVE COSTUME,

With their Canoe, Hut, Bows and Arrows, &c.

From the Manchester Guardian, Jan. 5, 1848.

THE ESQUIMAUX.—Yesterday, we visited in the lecture-theatre of the Mechanic's Institution, one of those outlying varieties of the human family, not often seen in this country,—a young male and female Esquimaux, natives of Cumberland, on the south-west coast of Davis's Straits, in 65° 20 north latitude, and 67° west longitude. They were brought to this country by Captain Parker (of the whaling ship Truelove, of Hull), who, after having made upwards of twenty voyages to that coast, has had his sympathies so much awakened for a people perishing of hunger, that he has brought this couple hither, in order to bring the condition of the tribes throughout the west coast of Davis' Straits (which is British territory) under the notice of our people and government. It seems that while similar tribes of people along the whole of the east coast, or East Greenland, are living in comfort and plenty, under Danish rule, supplied by the Danes with implements of the chase and the fishery, and as happy as external circumstances can make them, the wretched people on the opposite coast of Baffin's Bay—speaking a dialect of the same language, and being to all appearance the same people—are in the most destitute condition; and that chiefly from want of fire-arms and other means of getting food by the chase on sea and land. Thousands of the wretched denizens of British territory on the west side of the bay, are now dependent on the charity of the captains and crews of whaling vessels, for the means of existence; and several of these captains, we learn, distribute amongst these poor polar savages, large quantities of food and clothing every voyage. Captain Penny of Aberdeen, nearly emptied his own clothes' chest to clothe them; and Captain Parker has expended upwards of £30 in procuring necessaries for them. The history of the two poor young creatures now here, is brief, but striking. Betrothed, as is the custom of the country, when children of four or five years of age, Memiadluk, the husband, is now only 17, and Uckaluk, the wife, only 15 years old. On the Truelove reaching the coast, Uckaluk, having just lost her mother, and being thus left an orphan, and without the means of subsistence (as all the possessions of the deceased are buried in the same grave) had nothing before her but to live as the dogs do, and perhaps to be devoured by them. Won by Captain Parker's long-tried kindness to the natives, she implored him to take her to England. He refused to take her alone, or to take a male and female, unless married. The two betrothed accordingly became man and wife the night before the ship sailed (the male being shorn of his long hair as a part of the ceremony), and were then received, together with Memiadluk's canoe, spear, small tent of skins, &c., on board the ship, and brought to this country. They were in the most filthy state, their skin coated with oil or grease, and covered with vermin, and, the girl especially much emaciated by want of food. They were cleansed, and new sealskin garments given them, and they are now cleanly in habit, washing once or twice every day. They are, even to each other, taciturn; occasionally subject to great depression; but gentle, docile, grateful, and evidently much attached to Captain Parker. Both suffered considerably at first from sea-sickness, and having experienced every attention from Captain Parker, and Mr. Gedney, the ship's surgeon, Uckaluk went to the former with tears in her eyes, and said—"Uckaluk no father, no mother; Captain Parker be her father, doctor be her mother." In their own country these people eat all their food raw, and will devour from 7lb to 8lb or 9lb of flesh daily; and during the voyage they ate quantities of raw leg of beef; but on reaching England they soon learned to eat cooked meat, and though taking 2lb or 3lb at first, their appetites are now not much greater than those of healthy labourers in this country. Being forewarned by Captain Parker, they never touch intoxicating liquors, and their only beverage is cold water. On recovering from sea-sickness, Memiadluk made himself useful on board, helping the sailors in various ways; and on reaching Hull, Uckaluk was taken to Captain Parker's house, and there soon learned to wash clothes, glass and crockery, clean knives and forks, &c. The exhibition is throughout a simple, but interesting one. Both male and female are clothed in a neat dress of sealskin; their stature is low, their colour dark, like that of the quadroon, but with long flowing black hair; their features seem a mixture of the Malay and the African, and in mild, sad expression, resemble the Hottentot. The male gets into his canoe, holds the paddle and poises the spear. A description, from which we have gleaned the above particulars, is given by Mr. Gedney, surgeon of the Truelove; and afterwards Captain Parker explains the condition of the people on both sides of Baffin's Bay, and draws that contrast between the Danish and the British rule, which is so little to the credit of our country. It appears that muskets and ammunition are all that are wanting to place these poor creatures in a condition of comparative comfort; and that the Danes, by doing this, by sending them medical men and missionaries, and building them wooden cabins, have not only increased their comforts, but succeeded in establishing a lucrative trade, taking from them whale and seal oil, and the skins of the seal, the bear, the fox, &c., and sending them guns and ammunition in return. Captain Parker is seeking by this exhibition to induce the British government to pursue some such kindly policy towards these territorial subjects of Queen Victoria; and as to the two Esquimaux, he is collecting for them with the proceeds of their exhibition, a good stock of these and other necessaries of Esquimaux life, preparatory to their return; and he assures us he shall take them back on his next whaling voyage, leaving Hull about March next.

DAY EXHIBITION, from 2 to 4, ADMISSION, ONE SHILLING
EVENING EXHIBITION at Seven o'Clock, ADMISSION, SIXPENCE.
SCHOOLS AND CHILDREN, HALF PRICE,

BROCKS AND LENG, PRINTERS, BOWLALLEY-LANE, HULL.

An 1848 poster announcing the exhibition of two Eskimos, called 'Yacks' by the whalemen.

turned first to the Greenland Sea, and then began to sail west, across the Atlantic and into Davis Straits.

Here the same heedless slaughter continued, with the ever-diminishing numbers of whales retreating further and further north, into the hazardous waters of Baffin Bay. Eventually, the whalemen were having to follow their 2,000 mile sail across the Atlantic from Shetland to Cape Farewell, by a journey 1,000 miles north, up the west coast of Greenland, back down the east coast of Baffin Island; a desolate, mountainous wilderness, 2½ times the size of Britain.

Journeys took longer and longer. In 1799, Captain Sadler, of Hull, brought back the *Molly* a full ship on June 11th, 87 days after sailing, but more and more vessels were having to forsake Greenland for Baffin Bay, and the season was stretching inexorably. Ships which sailed in February or March were now not expected back until October, and, in bad ice seasons, much later. The *Swan,* of Hull, was one of these, in the 1837 season.

In a really bad ice year whole areas of Baffin Bay are covered with pack ice, in vast and solid fields, and in ice which is interlaced with lanes and polynias of clear water. Often there are bergs, spawned in their yearly thousands by the glaciers of Greenland, splintering through the pack in whatever direction the deep-sea currents are pushing them. The *Swan,* whaling in the high north-west of Baffin Bay, and with several men from the wrecked *Margaret,* of London, aboard, found her escape south blocked, in late August, and became trapped, in latitude 74 degrees.

Captain Dring was, fortunately, an experienced and determined man —religious, as many of these whaling captains were—and instituted a disciplined regime to spread out provisions and fuel, and to keep the ship reasonably clean and tolerable to be in. By the beginning of October the temperature was already nine degrees below zero, and canvas awnings were spread along the length of the deck, with long tubes fixed to carry away the lamp smoke. The winter was long, and terrible, the thermometer mercury freezing at its lowest point of 38 degrees below zero. There was no sun for over ten weeks, and fuel was so low that the harpooners' stove was not allowed to be lit if the temperature was higher than 13 degrees below zero. Rations were barely above subsistence level. Scurvy became evident, in bleeding mouths, great lassitude, foul breath, and bruised, discoloured skin.

Unknown to the *Swan,* other vessels, from Dundee, Peterhead, Hull, Berwick, Shields, and Aberdeen, were also trapped, and in the January of 1838 the Treasury offered all expenses defrayed and rewards of £300 to each of the first five ships sailing from England or Scotland before February 5th with extra provisions; £500 for a ship rescued from the edge of the ice; and £1,000 for any ship rescued from within it.

On March 31st, the *Swan* had to be abandoned, temporarily, as it turned out, and 14 volunteers, nine Hull men and five Londoners, set off with a whale boat for a Danish settlement thought to be only five miles away. Already weak, the fog and snow and petrifying cold rapidly affected them, and three days later, struggling to get back, the survivors were now too weak to pull or launch the boat. Three men eventually reached the *Swan*, and of these, two died.

The *Swan* was eventually sighted on May 14th, by a Leith whaler, whose crew, it was later reported, refused to go to her aid, as they said that they were not paid to do so. A few days later three Dundee ships, and then two from Hull, sent men and provisions to the trapped vessel, whose crew was by now in a completely debilitated state, and got her out by cutting a channel through the thick ice to clear water. Twenty of her crew of 48 were dead; seven from Hull, two from Grimsby—one of whom, John Nuttal, had been married only three months before the voyage—and 11 Shetlanders, plus also four out of the five Londoners.

Amazingly, nobody in Hull knew anything about her until she appeared off the Yorkshire coast in early July, truly back from the dead, after a voyage which, scheduled for six months, had taken 15. Her welcome would no doubt be almost delirious, although it was said, one survivor's wife had remarried, being convinced that her husband would not return.

Footnote: Polynias are spaces of open water in the midst of ice.

Chapter Eight

THE 'DIANA'

YORKSHIRE whaling ended in 1869, with a North Sea storm wrecking the *Diana,* of Hull, on the Lincolnshire sands of Donna Nook, and the *Diana's* story is in some ways a reflection of the county's whaling, born in high hopes, having good seasons and bad, and ending in a state of deflation and uncertainty.

No ship has a better documented voyage than the third last of this Hull ship, thanks to a journal kept by the surgeon, Dr. Charles Edward Smith, of Kelvedon, in Essex. Dr. Smith was 29 years old when he joined the *Diana* in Humber Dock, on a cold, grey, day in February 1866; he was a non-graduate who had attended Edinburgh University, a Quaker by upbringing, a vigorous, passionate, cultured man, who on this trip was repelled by the deck hands—of the Shetlanders he was to write: 'These fellows are perpetually complaining of their chests and worry me for medicines. Their language is like the current of their lives, one perpetual drawl. In habit and person they are frightfully dirty . . . one Englishman is worth half a dozen such.' Yet he was, in his stiff and unbending way, harassed and worried by his own inadequacies, and those of his medicine box, disturbed when one of his verminous patients died, and increasingly conscious of the terrible difficulties which faced those least able to cope.

The ship was a three-masted barque, with a gaudy figurehead of Diana, the huntress, and was the first whaler in Britain to use steam, her screw propellor being operated by two great engines which gave her a maximum of 40 horse power. Even for that time, with marine engines still in their infancy, she was ridiculously underpowered.

In her first year under steam and sail she had killed 13,000 seals, and in 1861 she had killed so many whales that she returned a full ship. Yet increasingly she was competing against larger Scottish vessels, with more efficient engines. By 1866 she was one of the only two vessels remaining in Yorkshire's once proud whaling fleet, the other being the indestructible *Truelove,* which was still purely sail.

Diana's captain in 1866 was John Gravill, in his sixties, the oldest and most experienced captain in the English and Scottish fleets, a Hull man whose home was in Walker Street, a staunch Methodist who tried never to fish on the Sabbath, and a shareholder in the ship. Several of his crew had sailed with him for a number of years. His first mate was George Clarke, who signed on the crew list as chief harpooner, simply because he did not have the paper qualifications as mate; the captain of the tug which towed *Diana* down the Humber signed on as first mate for the sake of good order. There were two engineers and three firemen, a carpenter, a cooper, a cook, the surgeon, five harpooners, a steward, seven Hull seamen, and, later picked up at Lerwick, 26 Shetland men and boys.

Diana sailed north, past Iceland, and up past Jan Mayen into the high, ice-infested, latitudes, where Dr. Smith, who had been sea-sick most of the way, had to treat his first case of frostbite—one of the Shetland boys. By March the 27th the ship, with several others in the offing, was dodging the pack ice, and neither Dr. Smith, nor his rough-haired terrier, Gypsy, were enjoying themselves.

By Good Friday, March the 30th, the weather had deteriorated into a violent storm, with a tremendous swell, and with waves crashing the ice against the hull. George Clarke made the first of his many gloom-filled prophesies, that if they were saved it would be by a miracle. Captain Gravill advised the doctor to prepare in case they had to abandon ship, tackle was got ready to lower the boats, and the crew were employed incessantly on the pumps. The captain was a sick man, his body beginning to give way from the pounding it had taken through the years, and he and the doctor, with the wind screeching in the rigging and the pumps clanking monotonously, discussed death. Dr. Smith, who at college had courted popularity by making mock of his religion, 'a despiser and a scoffer,' began to wish that he had not.

The gale increased into a hurricane, the main staysail was torn to pieces, and the ship drifted rapidly south-east, under bare poles. Dr. Smith now found himself alternating between fits of depression, and a kind of fierce exhilaration, helping with the rope fenders and with the pumps, and bawling 'Rock of Ages' into the howling wind. Later, he wrote of the waves and ice-masses resembling the hills of Northumberland; of the constant roar of the wind being like an express train in a tunnel; and of the lowering grey sky looking like a funeral pall.

By the time the hurricane had blown itself out, the *Diana* had drifted 40 miles south of Jan Mayen. There was now a period of intense cold, the seas freezing instantly on the clothes of the crew as they strove to chop away ice from ropes and deck. George Clarke said that they could not have lasted more than four hours if they had abandoned ship for the ice; and had the ship been struck, she would have sunk like a stone, stern first, because of the hundreds of tons weight of engines and coals.

Back in Lerwick by April the 30th, with no seals, they heard that the *Polynia* and *Narwhal,* both of Dundee, had been badly damaged, that the *Windward,* of Dundee, was missing, and that the *Ranger,* a Norwegian ship, had sunk, with all hands drowned. To the doctor, land had never seemed so good, and 'the singing of the larks had never seemed so sweet.' With the *Diana* freshly coaled and provisioned, but held up by contrary winds, he and Bill Reynolds, a Hull harpooner married to a Shetland girl, wandered the treeless hills in fine weather.

On May the 8th, the wind changed, and, under steam and sail, and in the company of some of the other whalers which comprised the 17 strong English and Scottish fleet, the *Diana* sped across the Atlantic in nine days, an unprecedented time for her to cover a distance which, the previous year, had taken 11 weeks. Rounding Cape Farewell, she turned north and Dr. Smith saw his first icebergs, which looked huge to him, although the crew assured him that they were not.

The fleet pushed up into the Arctic Circle, keeping closely together, for mutual protection, there being no charts of the area other than those drawn up by the captains for their own use. Disko Island was passed on the 70th degree of latitude; there had once been a couplet which ran: 'With Disko dipping, You'll see whale fish skipping,' but that had been decades earlier, in the days of plenty.

The *Truelove* stayed by Disko and began to search the deep indentations of the coast, unwilling to follow the sail-steamers into the floating deserts of ice. *Diana* and the others continued to press north, following the precipitous coast.

There were vast permanent areas of drifting pack ice in these seas, the largest of which was the Middle Pack, which covered an area hundreds of miles long and scores and even hundreds of miles wide, from the top end of Baffin Bay into Davis Straits. Almost always, the whalers attempted to complete a circuit of it, by sailing up the west coast of Greenland, across the open water at the top, and down the east coast of Baffin Island, tracing an inverted letter 'U.'

In the far north-west, beyond 75 degrees of latitude, were the whaling grounds, but to reach them the ships had to cross the shifting ice-fields of Melville Bay. Here the Middle Pack often met up with the ice fringing the Greenland coast, and the whalers had to risk running through winding lanes which might terminate in a dead end, or which might close. Glaciers pushing down from the mountains calved into bergs, and the whole area was known as the 'Breaker's Yard.' In 1830, 21 ships had been crushed and destroyed here; more than the whole of the present fleet. This year, the *Diana* got through, but only after having been baulked twice by the ice, and at one stage with the crew putting provisions into the boats in case the ship had to be abandoned. Already her puny engines had given hints of trouble to come, for of all the ships, she was involved in the greatest difficulties, and the others had disappeared far to the west before she reached the open sea.

A few narwhal were caught and flensed; Eskimos came on board, for the men to trade with the officers and the women to dance long hours with the crew; Gypsy, who had been on heat in Shetland, had four live pups; and eventually, on June the 30th, Bill Reynolds' boat chased and harpooned a whale at four o'clock in the morning, and later the same day a second was caught, by Dick Byers. Dr. Smith calculated the value of the day's whaling at £2,050, but for the *Diana* the season's whaling was confined to this single day.

There was much foggy, cold weather, with strong easterly winds pushing down so much ice onto the ships that, by the end of July, the whole fleet was beset in a massive ice-field. The days dragged by frustratingly, with the weather continually cold and thick, although excitement was caused by a party of men appearing from out of the north, laboriously dragging two whale boats, and announcing themselves as part of the crew of the Peterhead whaler *Queen,* beset in an inlet since early October of the previous year, and given up as lost. The *Tay* and the *Intrepid,* of Dundee, and the *Diana,* were each asked to take a third of the 18 men until the *Queen* should get free, but Captain Gravill refused, giving as his reason that the stocks of food would be inadequate for his own crew if the *Diana* was at all delayed.

The decision caused friction with the other vessels; in defence, what the captain claimed was very true, and by now he was an ailing, worried, and thoroughly sick man.

The first signs of scurvy were apparent on one of the Dundee vessels by the second week of August, and, as soon as the ice slackened, so the dozen whalers forced a way south along the coast, leaving the *Queen* to help herself, for it would be winter within the month, and all the captains knew was that they would perhaps be in similar trouble if the easterly winds resumed.

This in fact happened, and the way south down the coastal strip was blocked. The weather was already deteriorating, with blizzards and dropping temperatures, and the nights becoming noticeably longer. One ship, the *Wildfire,* turned east to try and force through the Middle Pack, the remainder turned around to the north, to retrace their steps. As so often, the winds seemed to blow from every point of the compass, and the *Diana's* feeble engines were of little use. She was soon left so far behind, that, crossing easterly, she met the other ships coming back, having found that the pack across Melville Bay was absolutely solid, blocking a way down the coast of Greenland.

By the end of August, the *Diana,* having tailed back after the other whalers, was again a few miles off the coast of Baffin Island, at about 70 degrees of latitude, in company with only the *Intrepid,* of Dundee, Captain Deuchars, and the *Queen,* of Peterhead, now free at least temporarily, though sailing with a depleted and emaciated crew.

There were only a few miles of ice between the ships and open water, much of it new ice, only a few inches thick. Nevertheless, they could not get through. Unable to face the prospect of being trapped a second winter, the *Queen* stood away to the east, gambling that a narrow lane of clear water, meandering into the hazy distance, would not close.

On the first day of September, with the ice easing, both the *Intrepid* and the *Diana* got up steam, and the *Intrepid* forced her way through, though the Hull ship could not, despite several desperate attempts with whale boats endeavouring to pull the ship, and with men straining at the capstans to wind in ropes fixed to ice anchors. The *Intrepid,* whose engines were almost twice as powerful as those of the *Diana,* kept going, and by evening only her topsails were visible. The *Diana* was now completely alone. Ice began to close around her, and preparations were made in case the ship had to be abandoned.

She was now firmly embedded in the ice, and part of a south-drifting coastal current, which would eventually spew her out into open water, either whole or in pieces. Yet the *Diana* would have to drift several hundred miles, and in places the current seemed to move scarcely at all. On board there was little coal left, and food was limited to a small quantity of, mainly, hard ship's biscuit, and salt meat.

The *Queen,* on which there had only been one death, had shown that men could live through an Arctic winter, yet there had to be leadership and organisation. Now was the time, in the few weeks before iron-hard winter really set in, to augment food supplies by hunting; to decide what on the ship could be used as fuel; and to organise duties and responsibilities; yet Captain Gravill was suffering from asthma, dropsy, bronchitis, and nervous exhaustion, and George Clarke, the first mate, was no leader of men.

An unsuccessful attempt was made to reach a distant whale, dead on the ice; a bear was shot, and the meat given to the dogs; indiscriminate shooting of south-flying birds took place. Despair already had its fingers on the ship. Dr. Smith wrote at length about having to put trust in God, but prayer was not going to alter the climate of the region, nor increase the stocks of food.

Captain Gravill died at 7 a.m. on December the 26th, after there had been a horrific 3½ months of storms, intense cold, near wreck on the icebergs which approached implacably, grinding and crashing, and a temporary abandonment of the ship.

George Clarke was now the master of a drifting ship which had scarcely any coal left for the cook to use; and no heating at all in the crew's quarters, where ice formed a glittering sheet over every bit of woodwork. There was no tobacco left apart from that belonging to the officers, and no hot water for the seamen to make sugarless black tea—they attempted to boil their kettles on the oil lamps, but succeeded only in covering everything with clouds of soot. The cook augmented the coal by burning whale flesh, which gave out a fierce heat, but burned quickly and made a terrible stench. Scurvy was now apparent. On January the 8th, the first crewman, Fred Lockham, of Hull, died.

By mid-January the ship had been pushed by the tides into the vast mouth of Frobisher Bay, and here she remained for almost three months, moving many miles into the bay with each flood tide, but never far enough out on the ebb to get back into the south-drifting current.

In England, an approach had been made by the mayor and aldermen of Hull to the Admiralty for help, but the Admiralty replied that none knew where the *Diana* was, and even if they did, there was nothing anyone could do until spring.

On February 7th, Dr. Smith recorded his dinners for the week—breakfast and tea consisted only of ship's biscuit, of which each man and boy received 12 a week:

Sunday: ½ lb flour dumpling, ½ lb salt boiled beef.
Monday: Ladleful of weak pea soup, with biscuit crumbs, ½ lb salt pork.
Tuesday: Ladleful of oatmeal porridge, ½ lb salt beef.
Wednesday: Nothing.
Thursday: ½ lb salf beef.
Friday: Ladleful of pea soup, ½ lb salt pork.
Saturday: Ladleful of oatmeal porridge.

The officers had a little tea and coffee left, the crew, none. Only one dog was alive. Three men were dead, and most were affected by scurvy.

By the end of the month there was no tea left, no oatmeal, no pork, and the men were on the last cask of beef.

Dr. Smith's journal ended on March the 17th, the day of the final breakout after an epic struggle to get through a surging minefield of ice into the open sea beyond. His writing, formerly neat and ordered, was now as crabbed and unsteady as that of a very old man.

The ship was pushed by the prevailing westerly winds across the Atlantic, and drifted into Ronas Voe, on the west coast of Shetland, on April 2nd. That day, two Shetlandmen and a Hull man died.

Telegrams were sent from Lerwick to Hull, to Edinburgh, and to London, where *The Times* of April 9th, underneath a two line snippet that Her Majesty's staghounds would not throw off on Friday until after the arrival of the 10.30 train from London at High Wycombe, printed a single paragraph on the 'LONG LOST WHALER.' *The Scotsman* of April the 11th referred to the 'miserable, scurvy stricken, dysentry-worn men . . . a spectacle, once seen, never to be forgotten,' and said that the most pitiable sight of all was the ship's boys, 'with their young faces wearing a strange aged look not easily to be described.' Emmanual Webster, the first engineer, wrote home: 'Me and my cousin are first rate, and have been all the voyage. Our only complaint has been the knife and fork not being so brisk . . . my linnet is still living.'

Four Hull men and nine Shetlanders had died, and others were maimed by scurvy or frostbite. It was virtually the end of Hull's—and Yorkshire's whaling. Both the *Diana* and the *Truelove* went under the auctioneer's hammer in the November of 1867, and, although the *Diana* went whaling again in 1868, she brought back only three tons of seal oil and 11 Shetland ponies.

Chapter Nine

THE END OF THE TRADE

AS the 19th century progressed, and the Industrial Revolution in Britain increased its momentum, so the demand for oil grew, yet in Hull money was being invested less in whaling, and more in the crushing of vegetable seeds, mainly rape, for the oil content. Street and home lighting was increasing in towns and cities, yet although the advocates of whale oil claimed that it gave a brighter, and odourless, light, coal-gas was being used more and more extensively. Victorian crinolines used whalebone in quantity, but it was a passing fashion of more than normal nonsense, and, with open hearth fires, highly dangerous as well.

Before the War of Independence, the American whaling fleet had supplied a great deal of England's oil, and now, 50 years later, their revived, and huge, whaling fleet was doing the same again, providing oil of superior quality, for the New England sperm whalers 'tried out,' or boiled, the blubber on board ship whilst it was fresh. The British, on the other hand, stored the blubber to be boiled out on their return home, several months old. Figures in Customs records show that in 1802, the blubber boiled in the Greenland Yards of Hull had about a 70 per cent oil content, that year giving 2,467 tons of oil from 3,439 tons of blubber.

In England, the Southern Fishery for the sperm whale was conducted on a small scale mainly by ships from London and Bristol. No Whitby ships seem to have gone, and the *Hull Advertiser* of February the 28th, 1823, noted that only three ships appeared to have sailed from Hull, one in 1794, one in 1815, and one in 1821.

In 1818, 157 ships sailed from Britain to the Arctic, Hull sending 64, London 18, Aberdeen 14, Whitby and Peterhead 12 each, and other ports 37. By 1830, Hull was down to 33 ships, Whitby to two; and by 1840, Whitby had finished, and Hull had only four. There was a slight revival, including an experiment with iron-hulled whalers, but it was all over by 1869.

At various times during the history of British whaling the Government had made an outright yearly payment of a bounty to the owners, of between one and two pounds sterling per ton weight of the vessel, and this had often meant the difference between breaking even or suffering a substantial loss on a bad voyage. When the bounty, then standing at 30s per ton, finally finished in 1824, allied to increasing competition, then the writing was on the wall. Nevertheless, it took several catastrophic years to convince owners in both Whitby and Hull, and especially in the latter town, that ships and men could be better employed elsewhere.

1830 was such a year, for 35 ships left Yorkshire for the whaling and only 29 returned, with several of these being clean. Yet at least only a few men died; this was not so five years later.

On the 10th of October, 1835, the *Hull Rockingham* reported: 'We regret to state that the first news received of the whale fishery is of the most disheartening nature. Two noble vessels are complete wrecks in the ice, and the number of fish taken is so insignificant as to be scarcely worth mentioning.'

The two vessels were both of Hull, the *Isabella,* wrecked five months previously, in May, and the *Lee,* wrecked on July 4th: bad news for the owners; terrible news for wives and parents and children, uncertain as to who was alive, and who was dead. Whaling and coalmining had an affinity in that every year there seemed to be at least one major disaster, and the bad news was no less numbing through being half-expected. Now there was nothing to do but wait—and history made it certain that some would wait in vain.

As the weeks passed, so news began to be brought back by returning whalers at Hull and Whitby, at Newcastle, Kirkcaldy, Dundee, Stromness, and Lerwick. It was not good. The season had been notable for ice and for ugly weather. There had been much fog. Summer had come late, and departed early.

The *Mary Frances* of Hull was the third to go, crushed in the ice of Baffin Bay on August the 20th. The captain and crew were taken aboard the *Lady Jane,* of Newcastle, but men from both crews returned to try and salvage provisions off the sinking *Mary Frances,* and got at the rum casks, set fire to the ship, and became involved in a fight. The wilderness was treated to the sight of one ship trapped, another on fire, and a yelling, cursing mob brawling on the ice. One of the Hull men later composed a poem:

> 'The ships ran to eastward, left us and *Lady Jane,*
> We tried to follow after, but our efforts proved in vain,
> For the ice came down upon us, with such a dreadful crush,
> That it did break our sternpost, when the water in did rush.
> The ship was lifted on the ice, just by her seven-foot mark,
> We got our clothes upon the ice, before that it was dark,
> We hoist our ensign union down, a sorry sight indeed . . .
>
> And now the riot it began, ranting, roaring, drinking,
> Drunken men upon the ice, when the good ship she was sinking,
> Orkney, Shetlands, Shields, and Hull, all were mixed together,
> They tumbled madly all along, over one another . . .'

This was on August the 21st. Six months later the grossly overcrowded *Lady Jane* had still not reached home; and some of the seamen never would.

On October the 20th, the *Dordon,* of Hull, was lost. The crew were taken aboard the *Abram,* also of Hull, which now had 90 men on board, with provisions for half that number. Trying to get out, the

Abram herself became trapped in the ice of Hudson Straits.

The news reaching home was so bad, with reports of eight ships missing, that a subscription fund was begun by a meeting of several Members of Parliament, at the Mansion House, in London. The *Cove,* a naval ship, was ordered to go to the rescue. This was an empty gesture, as almost everyone knew, but some politicians seemed to have little idea of what it was all about. Five years previously one of the Hull Members of Parliament had made an impassioned speech in which he referred to the whalemen as 'guardians of their country and her most heroic defenders in the hour of danger,' and gave every impression that he thought that whales were caught in the North Sea.

Ships expected back in September or October continued to limp home unexpectedly, in December, and in January, and when H.M.S. *Cove* eventually managed to sail from Stromness, Orkney, on February 26th, just two ships were still unaccounted for, the *Lady Jane,* of Newcastle, and the *William Torr,* of Hull. On March the 18th the *Lady Jane,* in a terrible state, limped into Stromness, with 22 dead.

The *William Torr* never did return. Two reports were made concerning her, the first when an American ship at Greenock reported seeing, in an Atlantic gale, the drifting hulk of a ship which might have been the Hull whaler; and the second when, in September, a cask marked 'William Torr,' which had obviously been in the sea for some months, was picked up in the Bay of Biscay.

Of the Hull fleet of 23 ships, five had been wrecked, and the remaining 18 had caught only 31 whales between them. Of the Whitby fleet, now down to two ships, the *Phoenix* had caught six whales, the *Camden* four, but the sea and wind were nothing if not persistent, and the *Camden,* in a November gale, collided with another vessel outside Whitby harbour, and was extensively damaged.

The season of 1837 was one of total loss for Whitby, when the town's one remaining whaleship, the *Camden,* under Captain Armstrong, returned clean, and the North Riding owners decided to pull out. A few ships from Hull continued to seal and whale, occasionally having a good season, more often a mediocre or a poor one. The whole of English whaling was by now in the doldrums, although the Scots, by increasingly killing young whales, and by shifting emphasis on to sealing, continued to increase their fleet. This was despite such appalling disasters as that of the *Dee,* of Aberdeen, which, beset in 1836, was picked up off Lewis, in the April of 1837, a floating morgue with only nine of her 46 crew alive. Even worse, in June the *Advice,* of Dundee, drifted onto the west coast of Ireland with only seven alive out of 49.

Hull had the first screw auxiliary whaler, that is, steam and sail, with the *Diana* in 1857, and in 1859 the town sent 12 whalers, out of a total Scottish-English fleet of 63. This year several of the Hull ships were iron hulled, which, it was considered, would enable them better to resist the ice, and many vessels in the fleet were fitted with 'bomb lances,' which were a type of iron-cased rocket about a foot in length and weighing perhaps four or five pounds. It was thought that with these a whale could be killed within a few seconds, even at a distance of 100 yards. The *Hull News* of February the 5th reported: 'There is much enthusiasm for the whale fishery.'

However, there was considerably less a couple of months later, for the iron ships proved to be a disaster. Wood would give when hit by the ice, but iron cracked. Wood could be mended, or at least patched, but there were no yards which could repair iron hulls nearer to the Greenland sealing grounds than Aberdeen. Of the British fleet, three iron ships sank. Of the five Hull ones, one had reasonable success with 8,350 seals, one caught 1,206, one 900, one caught nothing and had to return, damaged, and one caught only five before losing its screw.

In 1865, with the number of fishing smacks operating from the town now reaching 2,700, Hull had only two whalers left, the *Diana* and the *Aeolus.* Both ships went sealing first and although the *Diana* had no success, she left for the whaling immediately after taking on fresh provisions at Shetland, and killed several whales in Baffin Bay.

The *Aeolus,* under Captain Gray, which had left Hull on February the 15th, and had taken on 30 Shetland crewmen at Lerwick, caught a few seals in the Greenland Sea, and then sighted a great body of them extending for several miles within the ice, lying close together, a thick, meandering roadway of the creatures. The ship pushed through the twisting lanes of open water within the pack to get at them, but the ice began to move, and cut off her escape, and nipped her, every timber in the ship groaning and screeching as if in agony—which indeed she was, for several planks were fractured, and when the ice eased and the ship settled, water spurted into the hold.

Captain Gray turned back for Iceland, and tried to keep the sea at bay by continuous pumping, which was an arduous operation requiring several men at a time to heave and slacken the ropes of the cumbersome wood and leather suction pumps. The Shetlanders, always fatalistic, and often in poor physical condition, increasingly wanted to give up as the pumps clanked monotonously, hour after hour, never seeming to gain on the water. The ship was heavy and sluggish, rolling and pitching in a fashion which made progress slow, movement laborious, and sleep difficult.

On the fifth day they reached the desolate north coast of Iceland, and the *Aeolus,* with her crew exhausted, was washed up on a shelving, volcanic-sand, beach, with a whale boat taking a rope ashore first through heavy breakers. As much was got off the ship as possible before she hammered herself to pieces, and the whale boats, which had cost £21 each to make in Hull, were sold to men from a local village for 3s each.

The crew now became involved in a tough march overland, lasting several days, to the nearest harbour where they might obtain sea-going transport. Following the coast, they received open-handed hospitality from the isolated Icelandic communities. Eventually they reached a fiord, where Captain Gray hired a small Danish smack. Laden down dangerously, it took them on a hazardous trip back to Shetland.

The end of whaling came when the *Diana* was wrecked at the mouth of the Humber, on October the 19th, 1869. She had, the newspapers said, intended to return in time for Hull Fair, on October 11th, but had again been trapped by ice, after killing one small whale, three walrus, two bears, and two seals. Had she not been delayed, she might have returned to berth uneventfully, instead of meeting one of the worst gales to hit the east coast in years. She was blown onto Donna Nook sands, where, although the crew were saved, she broke apart, and casks and spars and her huntress figurehead were strewn onto the beach.

The Hull Registrar of Shipping noted his records, in red ink:

'*Diana,* official number 12,705, lost
Donna Nook, 19/10/1869
Chief Registrar informed.'

Chapter Ten

THE END OF THE WHALES

YORKSHIRE'S whaling days ceased with the wrecking of the *Diana*, and this was also virtually the end of English interest in the Northern Fishery.

However, the Scots continued to fish the Arctic until into the present century, and when Norwegian whaling captain Svend Foyn perfected his explosive harpoon, and fixed it into the bows of small, but fast, steam-driven catchers, there was no longer any chance that the remaining whales might escape.

By the early years of this century the whalemen, and especially the Norwegians, were beginning to sweep through the Arctic like animated vacuum cleaners, intent on sucking out every last pound of flesh or fur which could be converted into money. As the large whales—Greenland, Humpback, Blue and Fin—became ever fewer, and sightings further apart, so the whalemen took to killing the smaller species. Seals were also slaughtered in great numbers, and, to a lesser extent, walrus, and killer whales. And as the Arctic seas became empty, so the hunters turned to the shores of Spitsbergen and Greenland, and decimated the reindeer, polar bears, foxes, and eider-ducks.

In the 1930s the Greenland Right whale was declared to be a protected species, perhaps too late for the animal to recover, for once a species falls below a certain number then it is doomed to extinction, and the Right whale may already be down to its final few survivors.

By this time English and Scottish factory ships, out of London and Leith, were catching more whales in the Antarctic in a couple of days than the combined Hull and Whitby fleets had caught in a whole season, and with far less risk to men or ships. Destruction of the world's largest creature proceeded at an appalling pace, interrupted only by the war, and in 1963 the blue whales had been thinned to such an extent that the last three factory ships of the British fleet were sold to the Japanese. What had begun at the end of the 16th century, with Hull and London squareriggers tentatively exploring the coasts of Spitsbergen, is now finishing in the temperate seas of the world, as the Japanese and the Russians, having whaled as ruthlessly and as efficiently in the Southern Hemisphere as ever the Norwegians did in the Arctic, hunt the rapidly diminishing stocks of the few species that are left, using vast factory ships, helicopters, and fast catchers with massive, blunt-headed harpoons which slam into the whale's body and there explode in a burst of shrapnel which ensures a prolonged, and agonising, death.

In his autobiography, Captain William Barron, of Hull, who had been whaling as late as the 1860s, wrote: 'At one time it was almost thought that the world would stand still if the supply of fur of seals, and bone and oil of whales, should cease. The supplies did cease, but still the world goes on.' Since 1869, there has been no further Yorkshire participation other than occasional machinery being supplied by Hull engineering firms, and successive generations in the county have had their practical knowledge of whales confined to beach strandings, and excursions of the smaller whales up the River Humber.

Before the Humber became heavily polluted, the smaller whales, especially the 30 foot, squid-eating, bottlenose whale, and the 18 foot, fish-eating, beluga, or white whale, penetrated high upstream—in 1905 a beluga was captured in the Ouse near Naburn Lock, a few miles from York.

Larger whales have visited the Yorkshire coast only involuntarily, to be stranded, generally by bad weather, and to die. The largest one recorded seems to have been the 76′–0″ long fin whale, washed ashore inside Spurn Point in 1892, and killed by the local inhabitants, 'stopping up the blowhole with seaweed, mud, and gravel'—an unnecessary act, for the great weight of a whale's body, deprived of the sea's buoyancy, presses onto its lungs and causes suffocation.

In 1907 a 30′–0″ minke whale, pregnant, with a four feet foetus, was washed ashore at Scalby Ness. In 1910 a 51′–0″ female fin whale was stranded, in March, at Cloughton Wyke, and a second, 69′–0″ long, in the same area later in the year. In 1921, a 27′–0″ minke came ashore in Scarborough's North Bay, and in 1936, one 28′–0″ long was washed ashore at Saltwick, near Whitby.

A newly born killer whale, of the type now popular in marine zoos, was washed ashore at Scarborough in 1927, and ten years later at Bridlington a 60′–0″ sperm whale came ashore on the south beach, during heavy winter seas. Local fishermen cut out its teeth with hacksaws, and the corporation presented a fellmonger with the swelling carcase, and asked him to get rid of it quickly.

Yorkshire's whaling days are gone for ever, although the narrow passages, whitewashed, pantiled cottages, and communal yards in Whitby still give an indication of the bustling whaling town it once was. Indeed, perhaps with imagination, the town is still recognisable as it was described over 200 years ago, in a letter from there to a York printer:

'The inhabitants, though mostly Sea-faring men, are of a mild, affable, Temper, and exceedingly courteous to Strangers . . . Whitby is at present an opulent Town, and abounds with rich and expert Mariners.'

The attractive, privately-owned, Whitby museum has a select collection of whaling exhibits and records. Even if nothing else had remained, the two-volume work on whaling and the Arctic published in 1820 by the younger William Scoresby, and still in print, is an unparalleled record of the contemporary trade. Much has been made of Captain Cook as a notable North Riding seaman; more should be made of the Scoresbies.

The city of Hull, as befits a place which for a time so dominated British whaling, now has an excellent maritime museum, with a comprehensive section devoted to whales and whaling, including several superb oil paintings by John Ward; and the local history section of the Central Library has a large collection on whaling, supervised by a knowledgeable and helpful staff.

In these records of the two whaling towns lies the memory of a tough life, in a callous trade, wherein lay so many ingredients of risk that one could reasonably ask why any man would be crazy enough to venture.

Harsh financial reasons were undoubtedly one answer; the spice of danger another.

ACKNOWLEDGMENTS

The illustrations are from the following sources: Hull Museums, pages 44-46, back cover; Hull Central Library—Local History Department, 42, 43; Brown Atkinson & Co. Ltd., Hull, 47. All other uncredited illustrations are courtesy of Hull Town Docks Museum.

Back cover illustration of the 'Truelove' from a watercolour by William Ward.